Computer Communications and Networks

Series editors

Jacek Rak, Department of Computer Communications, Faculty of Electronics, Telecommunications and Informatics, Gdansk University of Technology, Gdansk, Poland

A. J. Sammes, Cyber Security Centre, Faculty of Technology, De Montford University, Leicester, UK

The **Computer Communications and Networks** series is a range of textbooks, monographs and handbooks. It sets out to provide students, researchers, and non-specialists alike with a sure grounding in current knowledge, together with comprehensible access to the latest developments in computer communications and networking.

Emphasis is placed on clear and explanatory styles that support a tutorial approach, so that even the most complex of topics is presented in a lucid and intelligible manner.

More information about this series at http://www.springer.com/series/4198

Mohsen Jahanshahi · Fathollah Bistouni

Crossbar-Based Interconnection Networks

Blocking, Scalability, and Reliability

Springer

Mohsen Jahanshahi
Department of Computer Engineering,
  Central Tehran Branch
Islamic Azad University
Tehran
Iran

Fathollah Bistouni
Department of Computer Engineering,
  Central Tehran Branch
Islamic Azad University
Tehran
Iran

ISSN 1617-7975   ISSN 2197-8433   (electronic)
Computer Communications and Networks
ISBN 978-3-030-08707-4   ISBN 978-3-319-78473-1   (eBook)
https://doi.org/10.1007/978-3-319-78473-1

Printed on acid-free paper

This Springer imprint is published by the registered company Springer International Publishing AG part of Springer Nature
The registered company address is: Gewerbestrasse 11, 6330 Cham, Switzerland

Preface

Interconnection networks are used in multiprocessor systems in order to establish the connection between the various nodes such as processors and memory modules. Blocking problem has always been considered as one of the most challenging issues in these networks, which degrades network performance and consequently the performance of the whole system. In the meantime, the main option for dealing with this problem is the use of non-blocking crossbar networks. However, there are engineering and scaling difficulties to use these networks in large-scale systems; the number of pins on a VLSI chip cannot exceed a few hundreds, which restricts the size of the largest crossbar that should be integrated into a single VLSI chip. However, there is a reasonable solution to the exploitation of the crossbar network in large-scale systems. The solution is using small-size crossbar networks as building blocks for larger network sizes. So far, this idea has led to a variety of topologies designed to meet the problems of blocking and scalability. Therefore, having the knowledge of different types of these crossbar-based networks and their strengths and weaknesses is essential in the design and selection of efficient interconnection networks.

Based on the above discussions, the focus of this book is on the *Crossbar-Based Interconnection Networks* to provide different perspectives required to recognize these momentous networks. This book consists of the following chapters:

Chapter 1 (*Introduction*): This chapter contains some introductory information for understanding the role of interconnection networks in multiprocessor systems, blocking problem, crossbar network and scalability problem, and existing solutions to address the scalability and blocking problems.

Chapter 2 (*Interconnection Networks*): This chapter presents a classification of interconnection networks and then attempts to provide the necessary information to recognize any of the networks.

Chapter 3 (*Blocking Problem*): This chapter is focused on the problem of blocking and analysis of different existing solutions to solve the problems of blocking and scalability.

Chapter 4 (*Fault-Tolerant Multistage Interconnection Networks*): Use of fault-tolerant multistage interconnection networks is a scalable solution to the blocking problem reduction. Therefore, this chapter analyzes a variety of different approaches to improve fault tolerance on multistage interconnection networks.

Chapter 5 (*Scalable Crossbar Network*): This chapter discusses the scalable crossbar network, which is a non-blocking interconnection network and uses small-size crossbar switches as switching elements.

Tehran, Iran   Mohsen Jahanshahi
                                                                                   Fathollah Bistouni

Contents

Chapter 1
Introduction

1.1 High Computational Power

The need for high computational power has always existed. A portion of this need can be met by increasing CPU performance, which doubles approximately every three years [1, 2]. However, some new problems have emerged that solving them requires a great deal of computing power. The United States Government Office of Science and Technology Policy has determined some of these problems and challenges in science and technology. Solving these grand challenges will be made possible by a high computing power that currently is difficult to achieve. Some of these problems are as follows:

Computational Fluid Dynamics global climate change, long-range weather forecast, enhanced oil and gas recovery, computational ocean sciences, automotive and supersonic aircraft design, nuclear reactor design, and quiet submarines.

Electronic Structure Calculations for Designing New Materials immunological agents, chemical catalysts, human genome, drug design, semiconductors, and superconductors.

Plasma Dynamics for Fusion Energy Applications and Military combustion systems, nuclear fusion, air, sea, and undersea surveillance for safe and efficient military technology.

Calculations to Understand the Fundamentals of Matter astrophysics, quantum chromodynamics, seismology, structural analysis, and condensed matter theory.

Symbolic Computational natural language processing, speech recognition, computer vision, automated reasoning, image processing, data mining for modeling business and financial processes, and discrete and continuous simulations of design, manufacturing, and production issues like transportation systems.

Scientists will be able to model such complex systems in sufficient detail if the computing power improves. In addition, many fundamental and applied sciences

© Springer International Publishing AG, part of Springer Nature 2018
M. Jahanshahi and F. Bistouni, *Crossbar-Based Interconnection Networks*,
Computer Communications and Networks,
https://doi.org/10.1007/978-3-319-78473-1_1

need to speed up progress. That is why the Office of Science and Technology Policy (OSTP) has admitted that efforts focused on high computing power are a non-negligible need. In this regard, the High-End Computing Revitalization Task Force (HECRTF) has begun its mission under the National Science and Technology Council (NSTC). The main mission of HECRTF is the design and development of a high-end computing project to maintain US leadership in science and technology. In this project, the HECRTF urged prominent scientists who take advantage of high computational power to advance their research in different disciplines, to do two things: First, identify the major challenges that require high computing power; second, estimating the additional computing power that is required to achieve the goals. These estimates suggest that achieving goals requires a computing power that is 100–1000 times more powerful than current computing power [1, 3, 4].

Today's supercomputers such as the Japan's Earth Simulator built in 2002 by Nippon Electric Company (NEC) are able to provide about 36 trillion calculations per second. However, such speeds are still not satisfactory for some scientific challenges. This reflects the fact that the need for progress in the areas of hardware, software, and system technologies is still felt [4].

Trends in fast networks, distributed systems, and multiprocessor computers during the past decade show that parallelism is a perfect solution for high computational power [5]. Parallel processors can be defined as a computer system that is made of multiple processors that are linked together by an interconnection network and the software required for the management of processors working together [6]. The goal of such systems is to achieve a stronger computing power. In other words, it is expected that a multiprocessor system can achieve a faster computing power than the fastest single processor system. In addition, the use of a multiprocessor system can be more economical than using a high-performance single processor system. Another issue is about the fault-tolerant capability of multiprocessors against single points of failure in a single processor system. If one of the processors in a multiprocessor system is unable to provide the service, then the other processors can be used as alternatives for providing that service, although at a lower level of performance [4, 6, 7].

1.2 Interconnection Networks

There are some questions about the discussions taking place in the previous section: How processors can communicate and cooperate? How is the flow of data between the different processors? What type of interconnection is considered?

Various subsystems in a parallel computer such as processors, memories, and disks need a communication subsystem to connect with each other. Interconnection networks are used to meet this need. In general, an interconnection network is made up of a set of switches plus links for connecting the switches to each other. This structure can connect N input terminal to M output terminal that can be used for internal connections between processors, memory modules, and I/O devices.

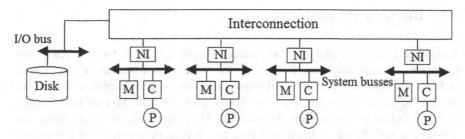

Fig. 1.1 Generic multiprocessor system with distributed memory

For a better understanding, consider a generic high-end parallel architecture. This system is shown in Fig. 1.1. Here, several processor nodes exist that an interconnection network provides the connection between them. This network is responsible for transferring data between the processor nodes. In this system, each node consists of three components: a (probably multicore) processor (*P*), a share of the main memory (*M*), and a cache hierarchy (*C*). In addition, the connection between processor nodes and global interconnection network can be created by a network interface (NI). I/O devices such as disks are also other important components of this system. I/O devices are often connected to an I/O bus, which is interfaced to the memory in each processor node via the interconnection.

Based on the above discussion, it can be argued that the three vital components of a multiprocessor computer are processor, memory hierarchy, and interconnection network. Therefore, multiprocessor systems require efficient interconnection networks in order to achieve a desired level of performance [1, 7–9]. In this regard, the design of an efficient network is a necessary phase.

The first step in designing an interconnection network is its topology selection. A key point in the performance of an interconnection network arises from the fact that communication resources are limited and shared. In other words, interconnection networks have been constructed with a set of shared switches connected to each other by a set of shared links, instead of creating a dedicated channel between each source–destination pair. With this view, the network's topology can be defined as the connection pattern used between the switching elements. In general, a network topology can be determined by a set of nodes *N* connected by a set of links *L*. Messages are sent by a subset of nodes called source terminal nodes *S* and are received by a subset of nodes called destination terminal nodes *D* where (*S* and *D*) ⊆ *N*. Therefore, a node on the network can be a terminal node that sends or receives packets, or it can be as a switch node that forwards the packets from input ports to selected output ports. As a result, a message for transmission from the source terminal node to a destination terminal node needs to pass some hops across the shared links and switching nodes [1, 10, 11].

1.3　Blocking Problem

A message is often divided into some packets before sending it, for an efficient and fair use of network resources. Packets are the smallest unit of data transfer, which includes the destination address and sequencing information, located in the packet header. In most topologies, packets should pass through several intermediate switches before reaching their destinations. Here, the task of determining the appropriate path that a packet should pass to reach its destination is the responsibility of routing algorithm. In each intermediate switches, routing algorithm selects the next channel that should be used. This channel can be selected from a set of candidate channels. Let us assume that all the selectable channels are busy, then the packet cannot continue in its path. As a result, the packet is blocked and this is known as the blocking or Head-of-Line (HOL) blocking problem. In fact, the main reason for this problem arises from the fact that network resources such as switches and links are limited and shared. In addition, efficient routing is essential for any network. When a packet arrives at a switch, a switching mechanism should determine when and how to connect the input channel to the output channel selected by the routing algorithm in that switch. In fact, the switching mechanism determines how to allocate the network resources to transmit messages [1, 10, 12–15].

In a more precise definition, a network is non-blocking if it can respond to all connection requests that are as a permutation of sources and destinations. Here, a "*permutation*" is defined as a request for simultaneous connections of every N sources to their N selected distinct destinations. Such network can be assigned a unique path to each source and its corresponding destination in each permutation without any conflicts (shared channels) with other source–destination pairs in that permutation. In contrast, if a network fails to respond to all possible requests configurable as permutation, without conflicts, then the network is blocking [1, 10, 16, 17]. As a result, blocking problem is a critical factor to consider in choosing an efficient interconnection topology.

1.4　Crossbar Network and Scalability Problem

So far, many interconnection topologies have been proposed. However, few of them are able to fix the blocking problem efficiently. Consider a system with N source and N destination nodes. In an ideal case, the sources and destinations can be connected to each other by a single $N \times N$ switching element. Such switches are known as crossbar network. A crossbar is able to connect any source node to each destination node in each permutation of the connections so that a large number of terminal nodes can be connected to each other simultaneously without conflict. Therefore, a crossbar network is strictly non-blocking. More precisely, the crossbar is defined as a network containing N sources (inputs) and M destinations (outputs), which makes it possible to min $\{N, M\}$ one-to-one interconnections without

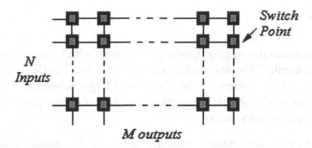

Fig. 1.2 A crossbar network of size $N \times M$

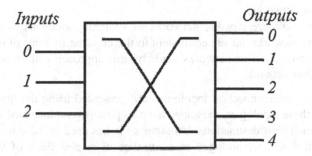

Fig. 1.3 Used symbol for a 3×5 crossbar network

contention. An $N \times M$ crossbar network is shown in Fig. 1.2. N is usually equal to M except for crossbars that are used to connect the processors and memory modules. It should be noted here that when we show a crossbar switch on a system, it is difficult to draw out all its details at any time. Therefore, a simple symbol commonly is used to it as shown in Fig. 1.3. In addition, we will omit the "X" and show the crossbar as a simple box with inputs and outputs.

Many other networks cannot achieve non-blocking mode that has been achieved easily in crossbar network. Therefore, is it possible to claim that the blocking problem is solved by crossbar without any trouble? The answer is that there is a scalability problem with crossbar. The advent of very large-scale integration (VLSI) has allowed integrating hardware required to crossbar switch in a single chip. However, the number of pins in a single VLSI chip cannot exceed a few hundreds. On the other hand, whatever the size of the network increases, the number of pins needed on the chip also increases. As a result, the number of available pins and wiring area are constraints that result in restriction on the size of the largest crossbar that can be implemented by a single chip [1, 10, 18]. This scalability problem prevents the use of crossbar networks in systems with large sizes. Therefore, crossbar network is usable in small-size multiprocessors.

1.5 Solutions

There is a way to use the crossbar network in systems with large size and solve the problem of scalability. The reasonable solution is to use small-size crossbar networks as the switching elements in the structure of larger networks. With a detailed look at pervious works, it can be inferred that two different approaches can be considered for using this solution:

(1) Using the small-size crossbar networks as switching elements in large-size interconnection networks that are different from the crossbar in terms of topology [1, 10, 16, 19–41]. This approach has been used in the design of many interconnection topologies, most of which are of the type of multistage inter-connection networks.
(2) Using the small-size crossbar networks as switching elements in larger inter-connection networks that are equivalent to the crossbar in terms of topology [1, 10, 42]. Interconnection networks made by this approach will be non-blocking like crossbar network.

So far, many interconnection topologies are presented using the first approach. Generally, in these topologies, messages have to pass several intermediate crossbar switches to reach their destinations. Crossbar switches used in these topologies are usually identical and are arranged in a number of stages. Each of these stages (except for the source–destination stages) has been connected to two stages located in before and after itself by regular connection patterns. Also, source (input) and destination (output) stages are connected to the first and last stages, respectively. These networks are recognized as multistage interconnection networks and depending on the number of stages and how to connect them to each other can have different characteristics. As will be discussed in Chap. 3, these networks can be an effective solution to solve the scalability problem. Another advantage of this type of networks that makes them attractive is cost-effectiveness. However, these networks cannot solve the blocking problem completely and effectively. One way to reduce the blocking problem in multistage interconnection networks is enhancing fault tolerance. Therefore, Chap. 4 is devoted to the study of some methods for improving fault tolerance in these networks.

Although many researchers have studied the first approach, less attention has been spent to take advantage of the second approach. However, efficient topologies can be designed by the second approach to provide important performance parameters in this area. For this reason, using this approach, and some of the ideas mentioned in [1, 10], a topology called Scalable Crossbar Network (SCN) [43] will be introduced in this book (see Chap. 5). SCN is a non-blocking network, and it can solve the scalability problem because of the use of small-size crossbar networks as the switching elements. In addition, SCN's routing mechanism is as self-routing, efficient, and affordable. Performance analysis conducted in Chap. 5 shows that SCN outperforms many MINs, namely Shuffle-Exchange Network (SEN), Extra-Stage Shuffle-Exchange Network (SEN+), Benes network, multilayer MINs, replicated MINs, and multistage implementation of crossbar network called Multistage Crossbar Network (MCN) in terms of various critical parameters such as terminal reliability, mean time to failure, and system failure rate. The representative networks will be discussed in Chap. 3.

References

1. Duato J, Yalamanchili S, Ni LM (2003) Interconnection networks: an engineering approach. Morgan Kaufmann, USA
2. Hennessy JL, Patterson DA (2012) Computer architecture: a quantitative approach. Elsevier, USA
3. Grammatikakis MD, Hsu DF, Kraetzl M (2000) Parallel system interconnections and communications. CRC Press, Boca Raton, Florida
4. Shiva SG (2006) Advanced computer architectures. CRC Press, Taylor and Francis Group
5. Jadhav SS (2009) Advanced computer architecture and computing. Technical Publications
6. El-Rewini H, Abd-El-Barr M (2005) Advanced computer architecture and parallel processing. Wiley
7. Dubois M, Annavaram M, Stenström P (2012) Parallel computer organization and design. Cambridge University Press
8. Culler DE, Singh JP, Gupta A (1999) Parallel computer architecture: a hardware/software approach. Morgan Kaufmann
9. Agrawal DP (1983) Graph theoretical analysis and design of multistage interconnection networks. IEEE Trans Comput 100(7):637–648
10. Dally WJ, Towles BP (2004) Principles and practices of interconnection networks. Morgan Kaufmann, San Francisco, Calif, USA
11. Wang X, Xiang D, Yu Z (2013) TM: a new and simple topology for interconnection networks. J Supercompu 66(1):514–538
12. Luo W, Xiang D (2012) An efficient adaptive deadlock-free routing algorithm for torus networks. IEEE Trans Parallel Distrib Syst 23(5):800–808
13. Garofalakis J, Stergiou E (2013) An analytical model for the performance evaluation of multistage interconnection networks with two class priorities. Future Gener Comput Syst 29 (1):114–129
14. Escudero-Sahuquillo J et al (2013) An effective and feasible congestion management technique for high-performance MINs with tag-based distributed routing. IEEE Trans Parallel Distrib Syst 24(10):1918–1929
15. Swaminathan K, Lakshminarayanan G, Ko S-B (2014) Design and verification of an efficient WISHBONE-based network interface for network on chip. Comput Electri Eng
16. Bistouni F, Jahanshahi M (2014) Improved extra group network: a new fault-tolerant multistage interconnection network. J Supercomput 69(1):161–199
17. Villar JA et al (2013) An integrated solution for QoS provision and congestion management in high-performance interconnection networks using deterministic source-based routing. J Supercomput 66(1):284–304
18. Hur JY et al (2007) Systematic customization of on-chip crossbar interconnects. In: Reconfigurable computing: architectures, tools and applications. Springer, Berlin, Heidelberg, pp 61–72
19. Bistouni F, Jahanshahi M (2015) Pars network: a multistage interconnection network with fault-tolerance capability. J Parallel Distrib Comput 75:168–183
20. Bistouni F, Jahanshahi M (2014) Analyzing the reliability of shuffle-exchange networks using reliability block diagrams. Reliab Eng Syst Safety 132:97–106

21. Parker DS, Raghavendra CS (1984) The gamma network. IEEE Trans Comput 100(4):367–373
22. Rajkumar S, Goyal Neeraj Kumar (2014) Design of 4-disjoint gamma interconnection network layouts and reliability analysis of gamma interconnection Networks. J Supercomput 69(1):468–491
23. Chen C-W, Chung C-P (2005) Designing a disjoint paths interconnection network with fault tolerance and collision solving. J Supercomput 34(1):63–80
24. Nitin, Garhwal S, Srivastava N (2011) Designing a fault-tolerant fully-chained combining switches multi-stage interconnection network with disjoint paths. J Supercomput 55(3):400–431
25. Wei S, Lee G (1988) Extra group network: a cost-effective fault-tolerant multistage interconnection network. In: ACM SIGARCH computer architecture news, vol 16. no 2. IEEE Computer Society Press
26. Matos D et al Hierarchical and multiple switching NoC with floorplan based adaptability. In: *Reconfigurable computing: architectures, tools and applications*. Springer, Berlin, Heidelberg, pp 179–184
27. Kumar VP, Reddy SM (1987) Augmented shuffle-exchange multistage interconnection networks. Computer 20(6):30–40
28. Vasiliadis DC, Rizos GE, Vassilakis C (2013) Modelling and performance study of finite-buffered blocking multistage interconnection networks supporting natively 2-class priority routing traffic. J Netw Comput Appl 36(2):723–737
29. Gunawan I (2008) Reliability analysis of shuffle-exchange network systems. Reliab Eng Syst Safety 93(2):271–276
30. Blake JT, Trivedi KS (1989) Reliability analysis of interconnection networks using hierarchical composition. IEEE Trans Reliab 38(1):111–120
31. Bansal PK, Joshi RC, Singh Kuldip (1994) On a fault-tolerant multistage interconnection network. Comput Electr Eng 20(4):335–345
32. Blake JT, Trivedi KS (1989) Multistage interconnection network reliability. IEEE Trans Comput 38(11):1600–1604
33. Nitin, Subramanian A (2008) Efficient algorithms and methods to solve dynamic MINs stability problem using stable matching with complete ties. J Discrete Algorithms 6(3):353–380
34. Fan CCh, Bruck J (2000) Tolerating multiple faults in multistage interconnection networks with minimal extra stages. IEEE Trans Comput 49(9):998–1004
35. Adams GB, Siegel HJ (1982) The extra stage cube: A fault-tolerant interconnection network for supersystems. IEEE Trans Comput 100(5):443–454
36. Tutsch D, Hommel G (2008) MLMIN: a multicore processor and parallel computer network topology for multicast. Comput Oper Res 35(12):3807–3821
37. Çam H (2001) Analysis of Shuffle-Exchange Networks under Permutation Traffic. In: Switching networks: recent advances. Springer US, pp 215–256
38. Çam Hasan (2003) Rearrangeability of (2n-1)-Stage Shuffle-Exchange Networks. SIAM J Comput 32(3):557–585
39. Dai H, Shen X (2008) Rearrangeability of 7-stage 16 × 16 shuffle exchange networks. Front Electr Electron Eng China 3(4):440–458
40. Beneš VE (1965) Mathematical theory of connecting networks and telephone traffic. vol 17. Academic Pr
41. Clos C (1953) A study of non-blocking switching networks. Bell Syst Tech J 32(2):406–424
42. Kolias C, Tomkos I (2005) Switch fabrics. IEEE Circuits Devices Mag 21(5):12–17
43. Bistouni F, Jahanshahi M (2015) Scalable crossbar network: a non-blocking interconnection network for large-scale systems. J Supercomput 71(2):697–728

Chapter 2
Interconnection Networks

2.1 Introduction

First of all, it should be noted that most of discussions in this chapter is based on Ref. [1].

In the previous chapter, it was argued that a multiprocessor system is made up of some processor units that are connected via some interconnection networks. In this system, the software is also a need to establish cooperation between different processors. This definition indicates that interconnection networks have an inalienable role in the performance of multiprocessor systems. Two important components must be considered in the classification of multiprocessor systems: processing units and interconnection networks used to connect them to each other. In addition, various communication styles can be considered for multiprocessor network. The communication styles can be divided based on communication model, namely shared-memory (single address space) or message passing (multiple address spaces). In a shared-memory system, communications are done via access to public memory by writing in or reading from it. However, communication within the message passing system is done by sending and receiving instructions. In any case, the important point is that interconnection networks are important in determining a connection speed in both systems. In this chapter, different types of interconnection topologies that are usable for connecting processors and memory modules will be examined. In general, two main types of interconnection networks will be discussed in this chapter: static interconnection networks and dynamic interconnection networks. Static networks create all the connections at the design stage. As a result, in these networks, messages should be sent via the preconstructed links and prefabricated paths. However, communication in a dynamic network follows an on-demand procedure. Dynamic network establishes the connection between two or more nodes, when sending messages over links. Some examples of static networks are networks of Cube, mesh, and ring. In addition, networks of bus, crossbar, and multistage interconnection networks are examples of dynamic interconnection networks.

© Springer International Publishing AG, part of Springer Nature 2018

M. Jahanshahi and F. Bistouni, *Crossbar-Based Interconnection Networks*,
Computer Communications and Networks,
https://doi.org/10.1007/978-3-319-78473-1_2

2.2   Classification of Interconnection Networks

In this section, a classification will be provided in terms of topology for inter-connection networks. Interconnection networks can be divided into two main classes: static and dynamic. Connections in static networks have been established by fixed links. However, connections in dynamic networks can be created dynamically on demand using switching components. In addition, static networks can also be divided into three groups: one-dimensional, two-dimensional, and hypercube. On the other hand, taking into account the different connection layout in dynamic networks, they can also be divided into two main groups: bus-based networks and switch-based networks. With a more detailed look, bus-based net-works, in turn, can be divided into two main types, namely single-bus network and multiple-bus networks. Finally, dynamic switch-based networks can include three different types of networks: single-stage, multistage, and crossbar. This classifica-tion is shown in Fig. 2.1. In the next sections, these different types of intercon-nection networks will be discussed in more detail.

2.3   Bus-Based Interconnection Networks

Bus network is an interconnection network topology with minimal complexity and popular among manufacturers [2–4]. In fact, the features that have led to the pop-ularity of the bus network can be expressed in three major aspects: low cost, easy implementation, and extensibility. Therefore, the bus network has been used in many computer and multiprocessor systems. Multimax and Alliant are examples of these multiprocessors [3]. Also, IBM RS/6000 R40, Sun Enterprise 6000, and HP 9000 K640 are some of the commercially available bus-based computers [1]. Furthermore, considering the positive characteristics of 3-D NoC (Network-on-Chip)-Bus Hybrid mesh structure such as power consumption, area, and performance, it is used in 3D-ICs (three-dimensional integrated circuits) [5–8]. Moreover, among the different

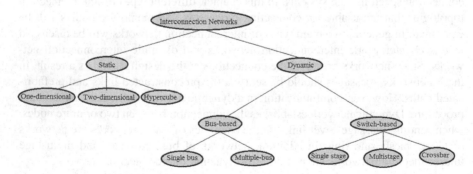

Fig. 2.1 A topology-based classification for interconnection networks

optical IN and NoC topologies, bus is quite promising due to its cost-effectiveness and simplicity for implementation and expansion [9–11].

Usually, bus networks are utilized for connecting the processors to memory modules. Nevertheless, in a typical bus network (i.e., single-bus network), a processor can access a memory module only when the requested module is not busy and bus is accessible. Actually, bus is not able to establish more than one connection between the processors and memories at the time. Consequently, one request can be replied at a time, and other requests are blocked. That is, if the number of processors is large, then there will be a long queue for the bus. In addition, the whole system will fail in case of bus failures for any reason. In fact, the bus network is a non-fault-tolerant network.

In fact, most of manufacturers prefer to use an interconnection network easy to implement and expand. Bus network can provide this feature. However, this network requires improving the reliability as well as fault tolerance. In this regard, one idea is the use of multiple buses in parallel. The multiple-bus networks provide several advantages: (1) They retain the strengths of a conventional bus such as ease of implementation and expansion of the bus network. (2) In case of unavailability of one of the buses, other ones can be utilized to make the connection. (3) They are fault-tolerant. The main conception to enhance fault-tolerance capability of interconnection networks is to build several paths from any source to any destination, possible to be provided via a multiple-bus network.

It is considerable that the approach of using multiple buses can be applied in a variety of forms: (1) multiple-bus with full bus-memory connection (MBF), (2) multiple-bus with single bus-memory connection (MBS), (3) multiple-bus with partial bus-memory connection (MBP), and (4) multiple-bus with class-based memory connection (MBC). These systems are described in detail in next subsections. One question arising here is which one of these schemes offer higher performance? So far, abundant literatures have been published on the performance of multiple-bus networks [1–4, 12–16]. However, some gaps exist in the previous works: (i) Most of previous works mainly focused on performance metrics of bandwidth and processor utilization. It should be mentioned that these parameters can present valuable information about the networks, but it has been proven that reliability is an undeniable need for all fault-tolerant systems [17]. However, in the past, this factor has received little attention. In addition, reliability analyses in the past are incomplete and partial. (ii) Most of previous works focused on the MBF and in some cases on MBP. However, MBS and MBC are also significant for evaluation. (iii) An expensive idea is not an acceptable solution practically. As a result, the next important metric is the cost-effective metric. Cost-effectiveness in the interconnection networks can be computed by dividing mean time to failure (MTTF) by hardware cost (hardware cost in the bus network equals to the number of its connections) [18–22]. To the best of our knowledge, this parameter has not been yet analyzed in multiple-bus networks domain.

Considering the above discussions, this section aims to give some ideas for filling the above-mentioned gaps [23].

In the sub-Sects. 2.3.1 and 2.3.2, the structures of single-bus network and multiple-bus networks will be discussed, respectively. Moreover, Sect. 2.4 provides significant performance analyses for bus-based networks.

2.3.1 Single-Bus Network

A single-bus network is the simplest structure to connect different nodes to each other (typically, processors and memory modules) on a multiprocessor system. A $P \times M$ single-bus network is shown in Fig. 2.2. Network complexity of bus is calculated by considering the number of used buses [1]. Therefore, network complexity of a single-bus network equals to $O(1)$. Although the bus network is simple and easy to expand, it is restricted considering that only one processor has access to the bus at a time. Furthermore, when the bus fails, for various possible reasons, the whole network fails.

Using multiple buses in a network is one solution to improve the single-bus network. In the multiple-bus networks, all processors are linked to all buses. Nevertheless, considering the connection method of memory modules to the buses, various connection patterns are possible which will be discussed in the next subsection.

2.3.2 Multiple-Bus Networks

In this subsection, various kinds of topologies in multiple-bus networks are explained.

A. *Multiple-bus with full bus-memory connection (MBF)*

A MBF of size $P \times M \times B$ is shown in Fig. 2.3. The basic idea in this network is that when larger numbers of memory modules are linked to more buses, more paths between any processor–memory pair can be attained. Therefore, in this structure, all memory modules are connected to all buses in the network. It is useful to define some basic features of multiple-bus networks here:

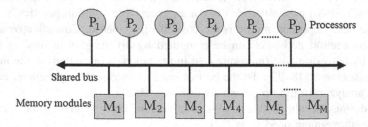

Fig. 2.2 A bus network of size $P \times M$

Definition 1 The number of connections in a multiple-bus network equals the number of links used to connect processors and memory modules to buses.

Definition 2 In a multibus network, load on each bus is defined by the number of links (connections) on the bus.

Definition 3 The term "permutation" is defined as a request for parallel connections of every P processors to their M corresponding distinct destinations.

The number of connections (i.e., C) and load on each bus (i.e., L) in a $P \times M \times B$ MBF are computed based on Eqs. (2.1) and (2.2), respectively, where B, P, and M denote the number of buses, processors, and memory modules:

$$C(MBF) = (B(P+M)) \qquad (2.1)$$

$$L(MBF) = (P+M) \qquad (2.2)$$

This network can provide several paths between each processor–memory pair. In addition, it can solve the blocking problem for each connection permutation. Nevertheless, the network requires many connections, and the load on each bus is relatively high.

B. *Multiple-bus with single bus-memory connection (MBS)*

Figure 2.4 shows a MBS of size $P \times M \times B$. The main purpose of this network is to minimize the number of connections in network. As a result, in this structure, each memory module is linked to a single bus. Consequently, the number of connections and load on bus i in a $P \times M \times B$ MBS are counted by Eqs. (2.3) and (2.4), respectively:

$$C(MBS) = (BP+M) \qquad (2.3)$$

$$L(MBS) = (P+M_i) \qquad (2.4)$$

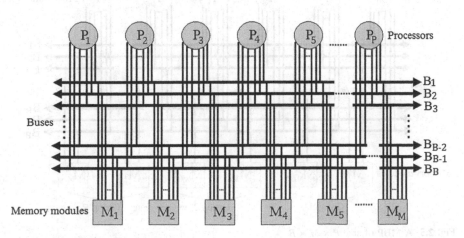

Fig. 2.3 A MBF of size $P \times M \times B$

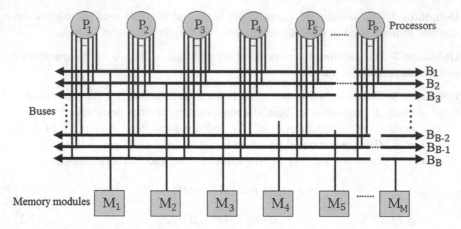

Fig. 2.4 A MBS of size $P \times M \times B$

where M_i is the number of memory modules connected to the bus i. This network has the least number of connections among the multiple-bus networks. Moreover, it is able to eliminate the blocking problem for all connection permutations. However, there is only one path between any processor–memory pair (i.e., non-fault-tolerant) in this network.

C. *Multiple-bus with partial bus-memory connection (MBP)*

A MBP of size $P \times M \times B$ is shown in Fig. 2.5. The purpose here is to build a middle number of connections comparing to the two previous topologies. In this topology, therefore, the memory modules are divided into g groups and each group of (M/g) memory modules is connected to a set of (B/g) buses. It is assumed that

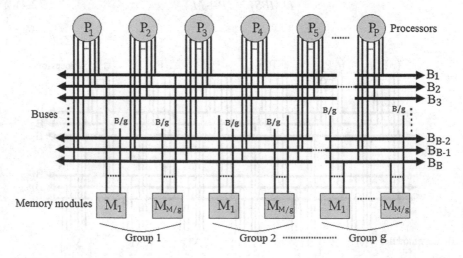

Fig. 2.5 A MBP of size $P \times M \times B$

g is a factor of both B and M. Usually, g is chosen so that connections are divided evenly on any bus. The number of connections and load on each bus in a $P \times M \times B$ MBP can be calculated via Eqs. (2.5) and (2.6), respectively.

$$C(\text{MBP}) = (B(P + (M/g)))\tag{2.5}$$

$$L(\text{MBP}) = P + (M/g)\tag{2.6}$$

This network is able to eliminate the blocking problem for any permutation of connections. Furthermore, it has smaller number of connections and is more reasonable than MBF. Although the number of paths between any processor–memory pair in this network is less than the MBF, it can supply (B/g) different paths between each processor–memory pair.

D. *Multiple-bus with class-based memory connection (MBC)*:

In Fig. 2.6, a MBC of size $P \times M \times B$ is shown. In a bus network, the number of connection requests to some of the memory modules may be more compared to other modules. To keep network performance at a desirable level, such memory modules should have larger number of connection buses. Hence, the core idea in MBC is to provide various levels of connections for various classes of memory modules. In MBC, memory modules are categorized in classes so that each class is connected to a different subset of buses. In general, the number of buses in each subset for a given class is different from the number of buses in the other ones. If the number of classes is k, then the memory modules in class k are connected to B buses from bus 1 to bus B and memory modules in class $k - 1$ are connected to $B - 1$ buses from bus 1 to bus $B - 1$. Generally, memory modules in class $j (1 \leq j \leq k)$ are linked to $(j + B - k)$ buses from bus 1 to bus $(j + B - k)$.

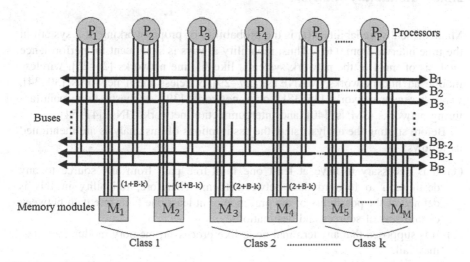

Fig. 2.6 A MBC of size $P \times M \times B$

Expectedly, subsets with larger number of buses are allocated to memory modules with higher traffic. The number of connections in a $P \times M \times B$ MBC is calculated by:

$$C(\text{MBC}) = (B \cdot P) + \sum_{j=1}^{k} M_j(j + B - k) \qquad (2.7)$$

where k is the number of classes and M_j shows the number of memory modules in class j. Furthermore, the load on bus i in the network is:

$$L(\text{MBC}) = \left(P + \sum_{j=\max(i+k-B,1)}^{k} M_j \right), \quad 1 \le i \le B \qquad (2.8)$$

This network is able to solve the blocking problem for each connection permutation. Moreover, the number of its connections is less than MBF, but more in comparison with MBS and usually more than MBP. This network can build multiple various paths between each processor–memory pair. However, the number of paths is different for various classes of memory modules.

Each of the above-mentioned multiple-bus networks has some weaknesses and strengths. Which network is more efficient? Some studies have been done in the past to answer this question. However, as it will be explained in the next subsection, reliability is one of the main metrics in fault-tolerant networks but has not been properly evaluated yet.

2.3.3 Reliability Analysis

Mathematically, reliability $R(t)$ is the probability of proper working of a system in the time interval from 0 to t. Thus, reliability always is significant for performance analysis of most of the network systems, like lifeline networks [24–26], wireless mobile ad hoc networks (MANETs) [27–29], wireless mesh networks [30–34], wireless sensor networks [35–38], social networks [39], stochastic-flow manufacturing networks (SMNs) [40], and interconnection networks (INs) [41–54].

Before starting the analysis step, the assumptions of this analysis are mentioned as follows:

(1) It is necessary to have at least one fault-free path from any source to any destination to forward data between them. Therefore, reliability in INs is defined as the possibility of the presence of at least one fault-free path between certain sets of sources and destinations.
(2) It is supposed that any terminal node like processor, memory module, and bus may fail.

(3) All failures are statistically independent.
(4) The failures are supposed to be exponentially distributed. As a result, we define λ_P, λ_M, and λ_B as the failure rate of processor, memory, and bus, respectively. Then, $R_P(t) = e^{-\lambda_P t}$, $R_M(t) = e^{-\lambda_M t}$, and $R_B(t) = e^{-\lambda_B t}$ give the corresponding reliabilities. Also, according to [14], it is assumed that a logical estimate for λ_P and λ_M equals 0.0001 per hour and a reasonable value for λ_B equals 0.00005 per hour.
(5) The network components have two statuses: working or failing.

We define network or all-terminal reliability as the probability of successful connection of all processors to all memory modules. Hence, in this case, from the reliability viewpoint all the processors and memory modules are crucial.

In a typical bus network, there are P processors, M memory modules, and a single bus. Therefore, the network reliability is calculated by:

$$R_{\text{Bus}}(t) = e^{-P\lambda_P t} e^{-M\lambda_M t} e^{-\lambda_B t} \qquad (2.9)$$

In Fig. 2.7, network RBD (reliability block diagram) for MBF has been shown. According to this diagram, we have:

$$R_{\text{MBF}}(t) = e^{-P\lambda_P t} e^{-M\lambda_M t} \left(1 - \left(1 - e^{-\lambda_B t}\right)^B\right) \qquad (2.10)$$

In the MBS, all of the processors, memory modules, and the buses are considered as a series system in terms of reliability. Thus, the network reliability equation for the MBS is given by:

$$R_{\text{MBS}}(t) = e^{-P\lambda_P t} e^{-M\lambda_M t} e^{-B\lambda_B t} \qquad (2.11)$$

Also, according to Fig. 2.8, we have:

$$R_{\text{MBP}}(t) = e^{-P\lambda_P t} e^{-M\lambda_M t} \left(1 - \left(1 - e^{-\lambda_B t}\right)^{\frac{B}{g}}\right)^g \qquad (2.12)$$

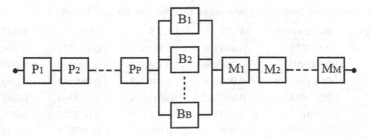

Fig. 2.7 Network RBD for $P \times M \times B$ MBF

In a similar manner, network RBD for MBC is illustrated in Fig. 2.9. Based on this diagram, network reliability for MBC is calculated by:

$$R_{\text{MBC}}(t) = \mathrm{e}^{-P\lambda_P t}\mathrm{e}^{-M\lambda_M t}\prod_{j=1}^{k}\left(1 - \left(1 - \mathrm{e}^{-\lambda_B t}\right)^{(j+B-k)}\right) \quad (2.13)$$

Based on Eqs. (2.9) through (2.13), the results of network reliability analysis as a function of time for network sizes $8 \times 8 \times 8$, $128 \times 128 \times 128$, and $512 \times 512 \times 512$ are summarized, respectively, in Tables 2.1, 2.2, and 2.3. In these results, for $B = 8$, k and $g = 4$; for $B = 128$, $k = 64$ and $g = 16$; and for $B = 512$, $k = 256$ and $g = 32$ are considered.

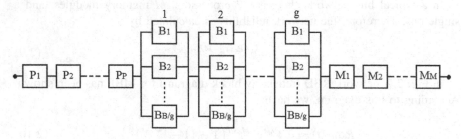

Fig. 2.8 Network RBD for $P \times M \times B$ MBP

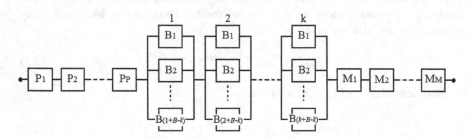

Fig. 2.9 Network RBD for $P \times M \times B$ MBC

Table 2.1 Network reliability as a function of time for network size $8 \times 8 \times 8$

Time (Hr)	Bus network	MBF	MBS	MBP	MBC
500	0.438235	0.449329	0.367879	0.448234	0.449329
1000	0.192050	0.201896	0.135335	0.199982	0.201896
1500	0.084163	0.090718	0.049787	0.088838	0.090718
2000	0.036883	0.040762	0.018316	0.039306	0.040762
2500	0.016163	0.018316	0.006738	0.017325	0.018315
3000	0.007083	0.008230	0.002479	0.007609	0.008229
3500	0.003104	0.003698	0.000912	0.003331	0.003697
4000	0.001360	0.001661	0.000335	0.001454	0.001661

Table 2.2 Network reliability as a function of time for network size $128 \times 128 \times 128$

Time (Hr)	Bus network	MBF	MBS	MBP	MBC
500	0.000003	0.000003	1.125352E−7	0.000003	0.000003
1000	7.250142E−12	7.621865E−12	1.266416E−14	7.621865E−12	7.621865E−12
1500	1.952180E−17	2.104224E−17	1.425164E−21	2.104224E−17	2.104224E−17
2000	5.256456E−23	5.809283E−23	1.603811E−28	5.809282E−23	5.809283E−23
2500	1.415358E−28	1.603811E−28	1.804851E−35	1.603810E−28	1.603811E−28
3000	3.811006E−34	4.427757E−34	2.031093E−42	4.427747E−34	4.427757E−34
3500	1.026155E−39	1.222403E−39	2.285694E−49	1.222394E−39	1.222403E−39
4000	2.763033E−45	3.374777E−45	2.572209E−56	3.374714E−45	3.374777E−45

Table 2.3 Network reliability as a function of time for network size $512 \times 512 \times 512$

Time (Hr)	Bus network	MBF	MBS	MBP	MBC
500	5.665851E−23	5.809283E−23	1.603811E−28	5.809283E−23	5.809283E−23
1000	3.210187E−45	3.374777E−45	2.572209E−56	3.374777E−45	3.374777E−45
1500	1.818844E−67	1.960503E−67	4.125337E−84	1.960503E−67	1.960503E−67
2000	1.030530E−89	1.138912E−89	6.616261E−112	1.138912E−89	1.138912E−89
2500	5.838830E−112	6.616261E−112	1.061123E−139	6.616261E−112	6.616261E−112
3000	3.308194E−134	3.843573E−134	1.701841E−167	3.843573E−134	3.843573E−134
3500	1.874374E−156	2.232840E−156	2.729431E−195	2.232840E−156	2.232840E−156
4000	1.061992E−178	1.297120E−178	4.377491E−223	1.297120E−178	1.297120E−178

According to Table 2.1, in terms of network reliability the best performance belongs to MBF and MBC. Moreover, the network reliability of the MBP is also very close to the two networks, especially in short-time operating. Furthermore, the weakest result belongs to the MBS. By increasing network size in Tables 2.2 and 2.3, network reliability of the three networks of MBF, MBC, and MBP comes closer to each other. In addition, the three networks always have higher network reliability than other ones. Consequently, the most efficient network can be chosen from three networks MBF, MBC, and MBP. Nevertheless, it should be considered that these networks have different number of connections. In fact, we have three networks with relatively close performances in terms of network reliability, but cost-wise they are different. As a result, it is to choose a network with lower cost compared to other networks; hence, MBP is a better option. However, for a detailed discussion, we ought to quantify this fact via evaluating the cost-effectiveness metric of point-network.

One can compute the cost-effectiveness of point-network equation for every network by the following equations:

$$CE_{\text{Bus}} = \int_0^\infty \left(e^{-P\lambda_P t} e^{-M\lambda_M t} e^{-\lambda_B t} \right) dt / P + M \tag{2.14}$$

$$CE_{\text{MBF}} = \int_0^\infty \left(e^{-P\lambda_P t} e^{-M\lambda_M t} \left(1 - \left(1 - e^{-\lambda_B t} \right)^B \right) \right) dt / (B(P+M)) \tag{2.15}$$

$$CE_{\text{MBS}} = \int_0^\infty \left(e^{-P\lambda_P t} e^{-M\lambda_M t} e^{-B\lambda_B t} \right) dt / (B.P+M) \tag{2.16}$$

$$CE_{\text{MBP}} = \int_0^\infty \left(e^{-P\lambda_P t} e^{-M\lambda_M t} \left(1 - \left(1 - e^{-\lambda_B t} \right)^{\frac{B}{g}} \right)^g \right) dt / \left(B \left(P + \left(\frac{M}{g} \right) \right) \right) \tag{2.17}$$

$$CE_{\text{MBC}} = \int_0^\infty \left(e^{-P\lambda_P t} e^{-M\lambda_M t} \prod_{j=1}^k \left(1 - \left(1 - e^{-\lambda_B t} \right)^{(j+B-k)} \right) \right) dt / \left((B.P) + \sum_{j=1}^k M_j (j+B-k) \right) \tag{2.18}$$

According to Eqs. (2.14) through (2.18), the results of cost-effectiveness of point-network analysis as a function of network size are illustrated in Fig. 2.10. As it is expected, the bus network has obtained the best result. However, it should be noted that this good result is due to its low cost, not because of its high performance. Therefore, although this network is attractive in terms of cost, it is not

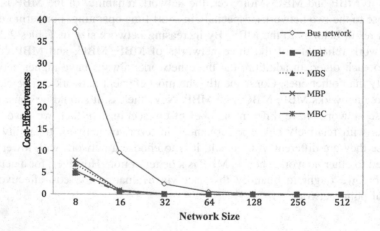

Fig. 2.10 Cost-effectiveness of point-network as a function of network size

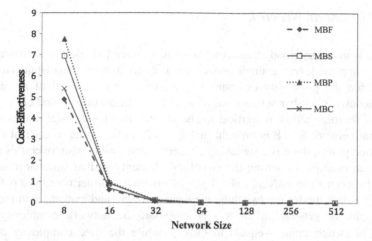

Fig. 2.11 Cost-effectiveness of point-network as a function of network size for multiple-bus networks

suitable in terms of reliability, fault tolerance, and blocking. That is why, researchers have been looking for better alternatives to this network. Based on reliability analyses in this section, multiple-bus networks can reasonably provide both performance and cost metrics. Therefore, we focus on these networks; Fig. 2.11 shows the performance of these networks more clearly in terms of cost-effectiveness of point-network. Considering Fig. 2.11, the MBP network displays best performance compared to other networks in terms of cost-effectiveness of point-network. The reason is that the MBP network has a great performance in terms of reliability considering the number of intra-connections compared to other multiple-bus topologies. Consequently, it is logical to select the MBP as the most cost-effective network in terms of network reliability.

In general, considering the analyses conducted in this section, we conclude that the MBP network has higher potential than other bus networks to meet two significant metrics of reliability and cost.

2.4   Switch-Based Interconnection Networks

In this type of networks, connection between processors and memory modules is built utilizing switching elements. There are three major interconnection topologies in this type: crossbar, single-stage, and multistage.

2.4.1 Crossbar Network

Crossbar is in a very good status comparing to a limited single-bus network. Bus network can provide only a single connection at a certain time; meanwhile, crossbar network has the possibility of connecting all of its inputs to all of its outputs simultaneously. Crossbar network has a switching element (also called cross point) at each of the intersections of vertical and horizontal lines inside itself. For example, a crossbar network 8×8 is brought in Fig. 2.12. In this case, in each of the 64 intersection points, there is a switching element (cross point). Moreover, this figure displays an example of setting the switching elements so that simultaneous connection between P_i and M_{8-i+1} (for $1 \leq i \leq 8$) is built. Furthermore, in this figure, the two possible modes for switching elements (straight and exchange) in crossbar are brought. In general, in an $N \times N$ crossbar, the network complexity—the number of switch points—equals to $O(N^2)$, while the time complexity of this

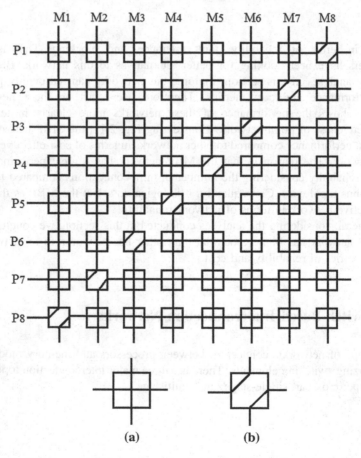

Fig. 2.12 An 8×8 crossbar (**a**) straight mode of switch and (**b**) exchange mode of switch

network—the delay from input to output—equals to $O(1)$. Moreover, it must be considered that simultaneous connection of multiple distinct input–output on the network in the form of a permutation should always be answered. Therefore, crossbar is a non-blocking network. However, the number of pins in VLSI chips cannot exceed a few hundred; hence, using large size crossbar networks that have been implemented in a single chip practically is not possible.

2.4.2 Single-Stage Network

Only one stage of switching elements exists between network inputs and outputs in this type. The simplest type of switch that can be used here is 2×2 crossbar switch. Figure 2.13 shows two possible states for the switch, namely straight and exchange. We have labeled the upper input and output lines by i and the lower input and output lines by j. (1) straight: input i to output i, input j to output j. (2) exchange: input i to output j, input j to output i.

Data circulate on the network for a time to create a connection between a specific input (source) and a specific output (destination). Some connection patterns exist for building the connection between input and output. One of the best-known examples to connect the network inputs and outputs in a single stage is Shuffle–Exchange. Two types of functions are done here: Shuffle and Exchange. These functions can be defined using a form of m-bit address of inputs: $p_{m-1}p_{m-2}\ldots p_1 p_0$, as follows:

$$S(p_{m-1}p_{m-2}\ldots p_1 p_0) = p_{m-2}p_{m-3}\ldots p_1 p_0 p_{m-1} \tag{2.19}$$

$$E(p_{m-1}p_{m-2}\ldots p_1 p_0) = p_{m-1}p_{m-2}\ldots p_1 \overline{p_0} \tag{2.20}$$

By functions of Shuffle (S) and Exchange (E), data will be in circulation from input to output until reaching the destination. If the number of inputs (e.g., processors) in a single-stage interconnection network equals to N and the number of outputs (e.g., memory modules) equals to N, then the number of switching elements in a stage equals to $\frac{N}{2}$. The maximum length of a path from an input to an output in the network, calculated by the number of switches along the path, equals to N. For

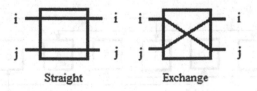

Straight   Exchange

Fig. 2.13 Two different settings to 2×2 switch

example, in a single-stage Shuffle–Exchange network of size 8×8, if the input is zero (000) and the destination is six (110), then the required Shuffle/Exchange functions and data flow will be as follows:

$$E(000) \rightarrow 1(001) \rightarrow S(001) \rightarrow 2(010) \rightarrow E(010) \rightarrow 3(011) \rightarrow S(011)$$
$$\rightarrow 6(110)$$

The network complexity and time complexity of single-stage network are both indicated by $O(N)$.

Moreover, there are some patterns of connections used in relationship formation between stages in interconnection networks. Among them, we can mention the cube network and Plus-Minus 2^i (PM2I) which are presented in below.

Connection pattern used in the cube network is defined as follows:

$$C_i(p_{m-1}p_{m-2}\cdots p_{i+1}p_i p_{i-1}\cdots p_1 p_0) = p_{m-1}p_{m-2}\cdots p_{i+1}\overline{p_i}\,p_{i-1}\cdots p_1 p_0 \qquad (2.21)$$

Considering a three-bit address $(N = 8)$, then we have: $C_2(6) = 2, C_1(7) = 5, C_0(4) = 5$. Figure 2.14 shows the cubic form of communication for a network with $N = 8$. This network is called Cube because it looks like connections between the corners of an n-dimensional cube $(n = \log_2 N)$.

PM2I network including $2k$ connection function is defined as follows:

$$PM2_{+i}(P) = P + 2^i \bmod N \ (0 \leq i < k) \qquad (2.22)$$

$$PM2_{-i}(P) = P - 2^i \bmod N \ (0 \leq i < k) \qquad (2.23)$$

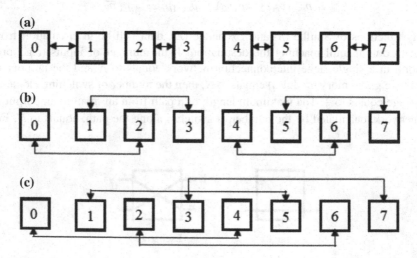

Fig. 2.14 Cube network for $N = 8$ (a) C_0, (b) C_1, (c) C_2

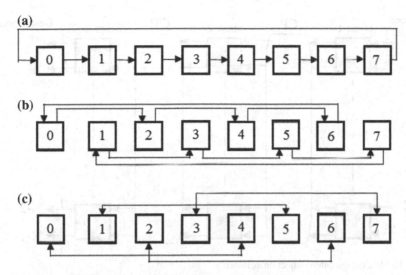

Fig. 2.15 PM2I network for $N = 8$ (a) $PM2_{+0}$ for $N = 8$, (b) $PM2_{+1}$ for $N = 8$, (c) $PM2_{+2}$ for $N = 8$

For instance, considering $N = 8$, then we have: $PM2_{+i}(4) = 4 + 2^i \bmod 8 = 6$. Figure 2.15 shows PM2I for $N = 8$. It should be noted that $PM2_{-(k-1)}(P) \forall P$, $0 \leq P < N$.

Butterfly network is another form of interconnection patterns defined as follows:

$$B(p_{m-1}p_{m-2}\cdots p_1 p_0) = p_0 p_{m-2}\cdots p_1 p_{m-1} \qquad (2.24)$$

As an example, considering a three-bit address ($N = 8$), butterfly mapping is provided below: $B(000) = 000$, $B(001) = 100$, $B(010) = 010$, $B(011) = 110$, $B(100) = 001$, $B(101) = 101$, $B(110) = 011$, $B(111) = 111$.

2.4.3   Multistage Networks

Multistage interconnection networks (MINs) were introduced to overcome some of single-bus system and crossbar network limitations. The most adverse limit of a single bus improved by MINs is related to availability of only one single path between the processors and memory modules. MIN is able to offer multiple paths between the processors and memory modules simultaneously. Moreover, these networks have the ability to solve the scalability problem on crossbar networks limited in implementing on a single chip. Of course, as discussed in the next chapter, these networks cannot be the perfect solution for the blocking problem. Blocking problem and the available solutions for this problem will be discussed in detail in the next chapter.

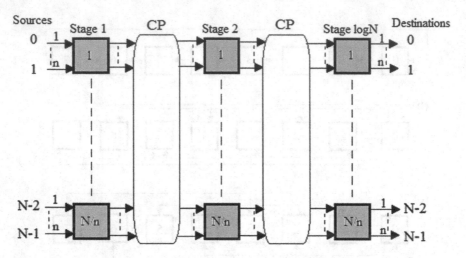

Fig. 2.16 Multistage interconnection network

As Fig. 2.16 shows, a multistage network interconnection generally includes a number of switching stages with a set of 2×2 switches on each stage. The stages are interconnected by a connection pattern between stages. These patterns may comply with any of the Shuffle–Exchange, Butterfly, Cube, etc.

As an example, Fig. 2.17 shows an 8×8 interconnection network using switches of size 2×2. This network is known as Shuffle–Exchange Network (SEN). The switch arrangement in this figure indicates method for creating a

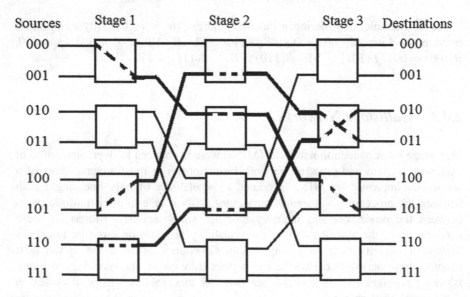

Fig. 2.17 SEN network of size 8×8

number of paths simultaneously on the network. For example, this figure shows establishment of three simultaneous paths to connect the three input–output pairs of $000 \rightarrow 101$, $101 \rightarrow 011$, and $110 \rightarrow 010$. It should be noted that the connection between stages follows the shuffle pattern.

In multistage interconnection networks, routing of a message from a given source to a given destination can be implemented on the basis of the destination address (this type of routing is known as self-routing). In an $N \times N$ MIN, $(\log_2 N)$ stages exist. Also, the number of bits in the destination address in this network is $(\log_2 N)$. Each bit in the destination address can be used for routing of messages in a specific stage. Destination address bits are tested from left to right and switching stages traveled from left to right based on these bits. The first bit is utilized to control the routing on the first stage; the next bit controls routing in the next stage, and so on. The contract used in message routing is in such a way that if the destination address bits for controlling the routing of a certain stage equals to zero, then the message is sent to upper output of switch. On the other hand, if the bit equals to one, then the message is sent to lower output of switch. For example, consider routing of a message from source 101 to destination 011 in a SEN of size 8×8 (shown in Fig. 2.17). Since the first bit of the destination address equals to zero, the message will be sent initially to an upper output of switch on the first stage. However, the next bit in the destination address equals to one; therefore, the message will be sent to lower output of switch in the middle stage. Finally, the last bit equals to one; consequently, the message is sent to lower output of switch in the last stage. This sequence leads the message to the desired output (shown in Fig. 2.17). As it can be seen, ease of message routings in the MINs is one of the desirable features of these networks. There are a number of other MINs including the Banyan network that is a known MIN. Figure 2.18 shows a Banyan network of size 8×8.

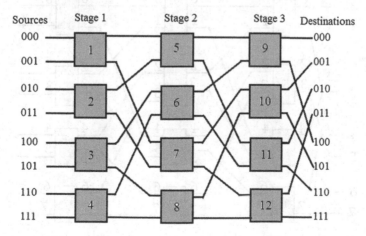

Fig. 2.18 Banyan network of size 8×8

Generally, if the number of inputs (sources), such as processors, as well as the number of outputs like memory modules in a MIN equals to N, then the number of stages is $(\log_2 N)$. Moreover, the number of switching elements in each stage is $\frac{N}{2}$. Consequently, network complexity that is measured in terms of the total number of switching elements equals to $O(N \log_2 N)$. In addition, the number of switches along the path is usually a measurement of delay a message encounters in routing from a source to a destination. Therefore, the time complexity, measured by the number of switching elements along the path from a source to the destination (i.e., path length), equals to $O(\log_2 N)$. For example, in a MIN of size 16×16, path length from a source to a destination equals to 4. Furthermore, the total number of switches in the network usually is a scale for the total network area. The total area of a MIN with size 16×16 equals to 32.

Omega network is another well-known MIN. An Omega network of size $N \times N$ includes $(\log_2 N)$ single-stage Shuffle–Exchange network. Each stage contains a column of $\frac{N}{2}$ number of two-input/output switches, all the inputs following a Shuffle connection pattern. Figure 2.19 shows an 8×8 Omega network. As can be seen in this figure, inputs at each stage use the Shuffle connection template. It should be noted that the connection in this network is the same connection used in the SEN network as shown in Fig. 2.17. However, the difference between the two networks is that the Omega network uses the Shuffle connection pattern even to connect the sources to the switches on the first stage.

A number of academic and commercial projects of MINs have been developed. From among, we can mention Texas Reconfigurable Array Computer (TRAC) at the University of Texas in Austin, Cedar at the University of Illinois, PR3 at IBM, Butterfly by BBN Laboratories, and NYU Ultracomputer at New York University.

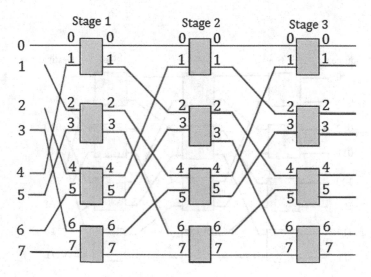

Fig. 2.19 Omega network of size 8×8

For instance, NYU Ultracomputer is a shared-memory MIMD architecture that can connect 4096 processors to 4096 memory modules via an Omega multistage interconnection network. In this system, the Omega network has improved in such a way that it is able to combine two or more requests for the same memory address. In addition, for managing the collisions of messages in the switch, the switch nodes at NYU Ultracomputer are supported with queues (queue lengths from 8 to 10 messages).

2.4.4 Blocking in Multistage Interconnection Networks

There are some classification criteria for MINs. However, blocking issue is one of the most important criteria. According to this criterion, the MINs can be classified as follows:

Blocking networks: Blocking networks have a feature that request for a new connection between an unused pair of input–output may be impossible due to a currently established connection between another input–output pair. Examples of blocking networks are Omega, Banyan, Shuffle–Exchange network (SEN), and Baseline. For example, consider the SEN network that is shown in Fig. 2.17. In the presence of a connection between the input 101 and output 011, the connection between input 100 and output 001 is not possible. This is because the connection of 101 to 011 uses upper output of the third switch from the top on the first stage, and this is the same output as needed for connecting 100 to 001. The contention will lead to network inability to satisfy the connection of 100 to 001; this problem is called blocking. However, it should be considered that if the connection of 101 to 011 is created, some requests for connection can be answered like 100 to 110.

Rearrangeable non-blocking networks: Rearrangeable non-blocking networks have a feature that can respond positively to all permutations that contain one-to-one and simultaneous connection requests. However, to build these connections, it may be required to reorganize some of the created connections. In this path reorganization, in the event of the unavailability of a path to a given source–destination, the rearrangeable network can take advantage of other available paths to connect by reorganizing paths. Consequently, this type of network can provide different paths between each source–destination pair. Benes network is one of the most well known of this type of networks. Figure 2.20 shows a Benes network of size 8×8. Two simultaneous connections are shown in this figure, in part (a). These two connections are $110 \rightarrow 100$ and $010 \rightarrow 110$. In this condition, consider a request to create a new connection of $101 \rightarrow 101$. However, we assume that in the presence of $110 \rightarrow 100$, connection of $101 \rightarrow 101$ is not possible due to busy status of other paths. In this case, one way to satisfy the new connection request is that connection of $110 \rightarrow 100$ be set again as shown in Fig. 2.20b.

Non-blocking networks: In non-blocking network, it is always possible to make connection between each unused pair of source–destination, even in the presence of other already-established connections between other source–destination pairs.

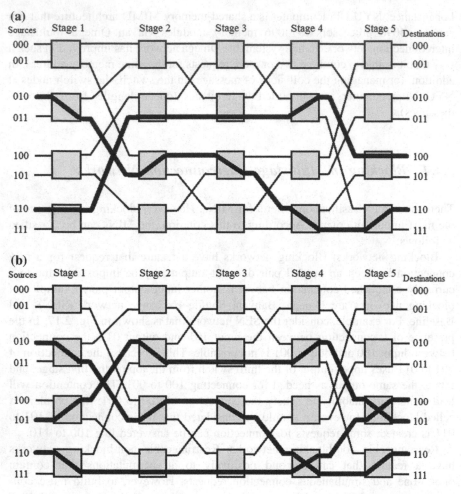

Fig. 2.20 **a** Benes network with two simultaneous connections and **b** rearrange of $110 \rightarrow 100$ in order to satisfy connection of $101 \rightarrow 101$

Clos network is the most popular non-blocking multistage network [55]. Clos network is a three-stage network in which each stage is composed of some crossbar switches. Clos networks with any odd number of stages can be derived from this three-stage network as recursive, so that middle-stage switches are replaced with the three-stage Clos network. Clos network can be described by a triple, (m, n, r). Here, m stands for the number of middle-stage switches, n indicates the number of input ports (output ports) in each switch in the first stage (last stage), and r shows the number of crossbar switches in each of first and last stages. In a Clos network, each switch in the middle stage has an incoming link from each switch in the first stage and it has an outgoing link to each switch in the last stage. Therefore, r is the number of first stage switches that are $n \times m$ crossbar switches and used to connect

n incoming ports to m middle-stage switches. Moreover, the m number of middle-stage switches are crossbars of size $r \times r$, which connect r switches of first stage to r switches of last stage. In addition, the r last-stage switches are also crossbar of size $m \times n$, which connect m middle-stage switches to n outgoing ports. For example, a Clos network represented by triple of (3,3,4) is shown in Fig. 2.21. A Clos network with triple of (5,3,4) is also shown in Fig. 2.22.

There is a remarkable point about Clos network; it can have non-blocking feature under some circumstances. For this case, the requirement is $m \geq 2n - 1$.

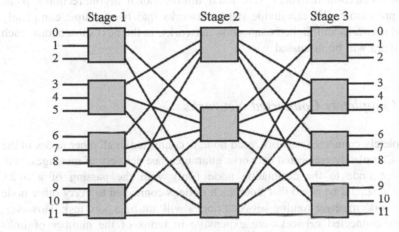

Fig. 2.21 A Clos network with triple of ($m = 3, n = 3, r = 4$)

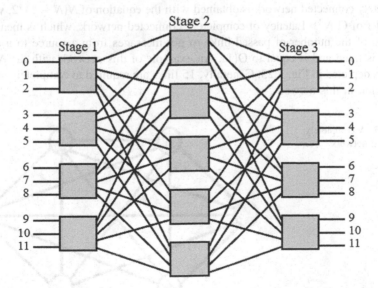

Fig. 2.22 A strictly non-blocking Clos network with triple of ($m = 5, n = 3, r = 4$)

According to this argument, Clos network shown in Fig. 2.21 is not non-blocking, but Clos network shown in Fig. 2.22 is non-blocking. On the other hand, it should be noted that if $m \geq n$, then Clos network is a rearrangeable non-blocking network. Therefore, the network shown in Fig. 2.21 is a rearrangeable non-blocking network.

2.5 Static Interconnection Networks

Static interconnection networks have fixed unidirectional or bidirectional paths between processors. We can divide static networks into two groups: completely connected networks and limited connection networks. In the next subsections, each of these types will be discussed.

2.5.1 Completely Connected Networks

In a completely connected network, each node is connected to all other nodes in the network. Completely connected networks guarantee fast delivery of messages from each source node to the destination node (only with the passing of a link). Moreover, it should be noted that since each node is connected to every other node of the network, message routing between nodes will not be a hard task. However, completely connected networks are expensive in terms of the number of links required for their construction. This disadvantage will be more and more apparent for large-size networks. It should be considered that the number of links in a completely connected network is obtained with the equation of $N(N-1)/2$, which has order of $O(N^2)$. Latency of completely connected network, which is measured in terms of the number of passed links to get messages from a source to a destination, is fixed and is equal to $O(1)$. An example of this network with the $N = 6$ nodes is depicted in Fig. 2.23. Generally, 15 links are required to completely satisfy connection in this network.

Fig. 2.23 A completely connected network

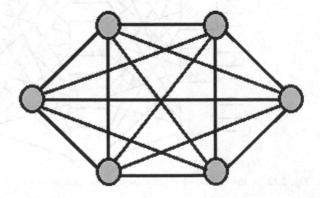

2.5.2   Limited Connection Networks

There would be no direct link from each node to every other node in the limited connection networks. Instead, the connections between some nodes are established via other nodes of the network. Therefore, the path length between nodes measured by the number of traveled links is expected to be longer than completely connected network. Moreover, two requirements are imposed due to limited connections in these networks: first, the requirement for interconnection pattern between nodes; second, the necessity of a mechanism for message routing within the network to deliver them to their destinations. These two subjects are further discussed below.

Over the years, some regular interconnection forms have been developed for limited connection networks. These connection patterns are:

- Linear arrays,
- Ring networks,
- Two-dimensional arrays,
- Tree networks,
- Cube networks.

Some simple examples of these networks are shown in Fig. 2.24.

In a linear array, each node is connected to two neighboring nodes. Two last nodes in the array are connected to a single neighbor. If the node i needs to connect to node j, and if $j > i$, then message in node i would have to travel nodes $i+1, i+2, \ldots, j-i$. But if $i > j$, then the message in node i must travel nodes $i-1, i-2, \ldots i-j$. In the worst-case scenario, when the node 1 sends a message to the node N, the message must pass the $N - 1$ nodes before reaching its destination. Therefore, although linear arrays have a simple architecture and a simple routing mechanism, they usually are slow. When the number of nodes (N) is large, this problem may appear. Actually, the network complexity of linear array is $O(N)$ as well as its time complexity is $O(N)$. If the two nodes at both ends of the linear array are connected to each other, then the resulting network will be ring architecture.

Binary tree (shown in Fig. 2.24d) is a special case of tree network. In this network, if a node in level i (it is assumed that the root level is zero) requires connection to a node at level j, where $i > j$ and the destination node belongs to the same root's child subtree, then the node should send the message to nodes on levels $i-1, i-2, \ldots, j+1$ to reach the destination node. If a node in level i needs connection to other nodes on the same level i (or with node at level $j \neq i$ where the destination node belongs to a different root's child subtree), it should send its message up the tree until the message reaches the root node at level 0. Afterward, the message should be sent down from the root nodes until it reaches its destination. It should be considered that the number of nodes (processors) in a binary tree with k levels can be calculated as follows:

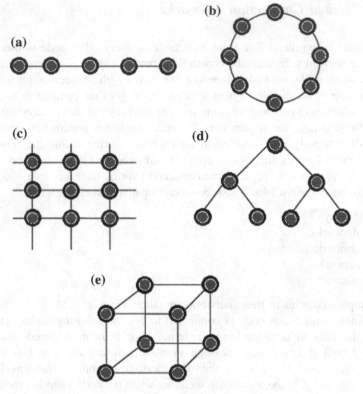

Fig. 2.24 Examples of static networks **a** a linear array network, **b** a ring network, **c** a two-dimensional array (mesh), **d** a tree network and **e** a 3-cube network

$$N(k) = 2^0 + 2^1 + 2^2 + \cdots + 2^k = \frac{(2^k - 1)}{2 - 1} = 2^k - 1 \qquad (2.25)$$

One should consider that the maximum depth of a binary tree system is $\lceil \log_2 N \rceil$. Therefore, network complexity is $\mathrm{O}(2^k)$, and time complexity is $\mathrm{O}(\log_2 N)$.

In this category of networks, researchers often focus on mesh and cube networks. These networks will be examined in detail in the next subsections.

2.5.3 Cube Networks

n-cubes (hypercubes/n-dimensional hypercubes) have many fine network characteristics like great connectivity, low density, small diameter, and regular structure. The reason is that the binary strings are naturally encoded into their topology which makes them global [56–58]. n-cubes are analyzed in different networking fields

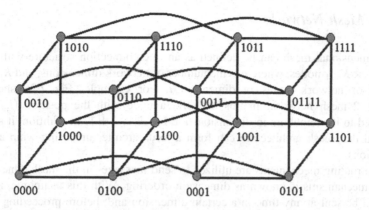

Fig. 2.25 Structure of a 4-cube

such as data center networks, interconnection networks, wavelength division multiplexing (WDM) networks [58–63]. The Intel Personal SuperComputer (Intel iPSC) is an example of commercial cube-based multiprocessor systems.

A n-cube is an undirected graph having 2^n vertices (nodes), and each vertex is connected to n other vertices. The vertices are labeled from 0 to $2^n - 1$ so that there is an edge between a certain pair of vertices if and only if binary representation of them differs in one and only one bit. This feature gives possibility to a straightforward routing mechanism. In this mechanism, the path from a source node i to destination node j can be achieved by XOR (exclusive OR) of binary addresses of nodes i and j. If the result of the XOR in a given bit equals to 1, then the dimension corresponding to the bit is the preferred route. A 4-cube is shown in Fig. 2.25 as an example. In this figure, consider the source node 0101 and destination node $1011 : 0101 \oplus 1011 = 1110$, and so the messages will be sent over the links corresponding to the dimensions 2, 3, and 4 (counting from right to left).

The Hamming distance between two addresses $S_n S_{n-1} \ldots S_1$ and $D_n D_{n-1} \ldots D_1$ in n-cube is given by:

$$NH(S, D) = \sum_{i=1}^{n} S_i \oplus D_i \qquad (2.26)$$

In a n-cube, between any pair of nodes S and D, there are $H(S, D)$ disjoint paths with length $H(S, D)$ and $n - H(S, D)$ disjoint paths with length $H(S, D) + 2$. For example, consider again the source 0101 and destination 1011 in Fig. 2.25: $H(S, D) = 3$, and therefore, three paths with length 3 and one path with length 5 exist between source and destination. Moreover, if the number of faulty links and nodes is less than n in a n-cube, then there is at least one path length less than or equal to $H(S, D) + 2$ between any two non-faulty nodes S and D.

The recursive nature of structure is one of the desired characteristics of n-cubes. This means that a n-cube can be constructed by connecting nodes with the same address in the two $n - 1$-cubes. For example, it should be noted that a 4-cube (shown in Fig. 2.25) is made up of two 3-cubes.

2.5.4 Mesh Networks

An n-dimensional mesh can be defined as an interconnection structure with $K_0 \times K_1 \times \cdots \times K_{n-1}$ nodes, where n is the number of network dimensions, and K_i is the number of network nodes in dimension i. For example, Fig. 2.26 shows a $3 \times 3 \times 2$ mesh network. In this network, a node with the position (i,j,k) is connected to its neighbors in dimensions $i \pm 1$, $j \pm 1$, $k \pm 1$. In addition, it should be noted that mesh architecture can form a torus architecture using wrap around connections.

Some routing mechanisms are utilized to send messages in the mesh. One of the routing mechanisms is known as dimension ordering. With this technique, a message will be sent at any time in a certain dimension and, before proceeding to the next dimension, enters with a perfect coordination in each dimension. As an example, consider a three-dimensional mesh. Since each node is identified by a location (i,j,k), messages are sent firstly over dimension i, then during the dimension j, and finally during dimension k. Figure 2.26 shows the path of a message from node S in the position $(0,0,0)$ to node D in position $(2,1,1)$. Some other routing mechanisms have been suggested in the mesh. The routing mechanisms include reversal routing, turn model routing, and node labeling routing. It should be considered that for a communication network between meshes with N nodes, the longest distance traveled between any two arbitrary nodes equals to $O(\sqrt{N})$. Multiprocessor systems with mesh interconnection networks are effectively able to support many scientific calculations. n-dimensional mesh can be built using only short wires and creating identical boards, each of which requires a small number of pins for connection to other boards. Another advantage of mesh networks is their scalability. Large-size meshes can be achieved without changing the

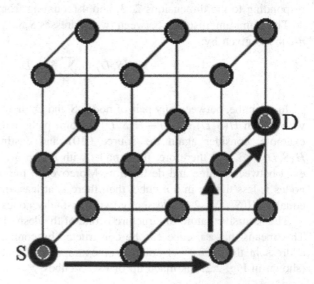

Fig. 2.26 A $3 \times 3 \times 2$ mesh network

degree of smaller mesh nodes (node degree is defined as the number of links connected to the node). As a result of these features, a large number of distributed memory parallel computers use the mesh interconnection networks. Some examples of these systems include the MPP from Goodyear Aerospace, Paragon from Intel, and J-Machine from MIT.

References

1. El-Rewini H, Abd-El-Barr M (2005) Advanced computer architecture and parallel processing. Wiley, Hoboken
2. De Almeida D, Kellert P (2000) Markovian and analytical models for multiple bus multiprocessor systems with memory blockings. J Syst Architect 46(5):455–477
3. Bhuyan LN, Yang Q, Agrawal DP (1989) Performance of multiprocessor interconnection networks. Computer 2:25–37
4. Akram S et al (2010) A workload-adaptive and reconfigurable bus architecture for multicore processors. Int J Reconfigurable Comput 2010:2
5. Rahmani A-M et al (2014) High-performance and fault-tolerant 3D NoC-bus hybrid architecture using ARB-NET-based adaptive monitoring platform. IEEE Trans Comput 63 (3):734–747
6. Yao C et al (2015) Low latency multicasting scheme for bufferless hybrid NoC-bus 3D on-chip networks. Comput Eng Technol 36–47 (Springer, Heidelberg)
7. Zheng J et al (2016) DFSB-based thermal management scheme for 3-D NoC-bus architectures. IEEE Trans Very Large Scale Integr VLSI Syst 24(3):920–931
8. Ebrahimi M et al (2013) Cluster-based topologies for 3D networks-on-chip using advanced inter-layer bus architecture. J Comput Syst Sci 79(4):475–491
9. Broadmeadow MAH, Walker GR, A LIN inspired optical bus for signal isolation in multilevel or modular power electronic converters. In: Proceedings of the IEEE 11th international conference on Power Electronics and Drive Systems (PEDS), pp 898–902
10. Zhang J, Yang X, Li X (2014) Wavelength assignment for locally twisted cube communication pattern on optical bus network-on-chip. Opt Fiber Technol 20(3):228–234
11. Bamiedakis N et al (2014) A 40 Gb/s optical bus for optical backplane interconnections. J Lightwave Technol 32(8):1526–1537
12. Irani KB, Önyüksel IH (1984) A closed-form solution for the performance analysis of multiple-bus multiprocessor systems. IEEE Trans Comput 100(11):1004–1012
13. Mudge TN et al (1986) Analysis of multiple-bus interconnection networks. J Parallel Distrib Comput 3(3):328–343
14. Das CR, Bhuyan LN (1985) Bandwidth availability of multiple-bus multiprocessors. IEEE Trans Comput 100(10):918–926
15. Yang Q, Zaky SG (1988) Communication performance in multiple-bus systems. IEEE Trans Comput 37(7):848–853
16. Yang Q, Bhuyan LN (1991) Analysis of packet-switched multiple-bus multiprocessor systems. IEEE Trans Comput 40(3):352–357
17. Koren I, Mani Krishna C (2007) Fault-tolerant systems. Morgan Kaufmann, USA
18. Bistouni F, Jahanshahi M (2014) Scalable crossbar network: a non-blocking interconnection network for large-scale systems. J Supercomputing 71(2):697–728
19. Jahanshahi M, Bistouni F (2015) Improving the reliability of the Benes network for use in large-scale systems. Microelectron Reliab 55(3):679–695
20. Bistouni F, Jahanshahi M (2015) Pars network: a multistage interconnection network with fault-tolerance capability. J Parallel Distrib Comput 75:168–183

21. Blake JT, Trivedi KS (1989) Reliability analysis of interconnection networks using hierarchical composition. IEEE Trans Reliab 38(1):111–120
22. Bansal PK, Joshi RC, Singh Kuldip (1994) On a fault-tolerant multistage interconnection network. Comput Electr Eng 20(4):335–345
23. Bistouni F, Jahanshahi M (2016) Reliability analysis of fault-tolerant bus-based interconnection networks. J Electron Test 32(5):541–568
24. Kang Won-Hee, Kliese Alyce (2014) A rapid reliability estimation method for directed acyclic lifeline networks with statistically dependent components. Reliab Eng Syst Safety 124:81–91
25. Liu H et al (2015) Vulnerability assessment and mitigation for the Chinese railway system under floods. Reliab Eng Syst Safety 137:58–68
26. Shuang Q, Zhang M, Yuan Y (2014) Node vulnerability of water distribution networks under cascading failures. Reliab Eng Syst Safety 124:132–141
27. Padmavathy N, Chaturvedi SK (2013) Evaluation of mobile ad hoc network reliability using propagation-based link reliability model. Reliab Eng Syst Safety 115:1–9
28. Meena KS, Vasanthi T (2016) Reliability analysis of mobile ad hoc networks using universal generating function. Qual Reliab Eng Int 32(1):111–122
29. Babaei H, Fathy M, Romoozi M (2014) Modeling and optimizing random walk content discovery protocol over mobile ad-hoc networks. Perform Eval 74:18–29
30. Jahanshahi M, Dehghan M, Meybodi MR (2013) LAMR: learning automata based multicast routing protocol for multi-channel multi-radio wireless mesh networks. Appl Intell 38(1):58–77
31. Jahanshahi M, Dehghan M, Meybodi MR (2013) On channel assignment and multicast routing in multi-channel multi-radio wireless mesh networks. Int J Ad Hoc Ubiquitous Comput 12(4):225–244
32. Chakraborty S, Nandi S (2014) Evaluating transport protocol performance over a wireless mesh backbone. Perform Eval 79:198–215
33. Jahanshahi M, Dehghan M, Meybodi MR (2011) A mathematical formulation for joint channel assignment and multicast routing in multi-channel multi-radio wireless mesh networks. J Netw Comput Appl 34(6):1869–1882
34. Jahanshahi M, Barmi AT (2014) Multicast routing protocols in wireless mesh networks: a survey. Computing 96(11):1029–1057
35. Jahanshahi M, Maddah M, Najafizadegan N (2013) Energy aware distributed partitioning detection and connectivity restoration algorithm in wireless sensor networks. Int J Math Model Comput 3(1):71–82
36. Wang C et al (2014) Reliability and lifetime modeling of wireless sensor nodes. Microelectron Reliab 54(1):160–166
37. Wang C et al (2016) Infrastructure communication sensitivity analysis of wireless sensor networks. Qual Reliab Eng Int 32(2):581–594
38. Jahanshahi M, Rahmani S, Ghaderi S et al (2013) An efficient cluster head selection algorithm for wireless sensor networks using fuzzy inference systems. Int J Smart Electr Eng (IJSEE) 2(2):121–125
39. Schneider K et al (2013) Social network analysis via multi-state reliability and conditional influence models. Reliab Eng Syst Safety 109:99–109
40. Lin Y-K, Chang P-C (2013) A novel reliability evaluation technique for stochastic-flow manufacturing networks with multiple production lines. IEEE Trans Reliab 62(1):92–104
41. Chang N-W et al (2015) Conditional diagnosability of (n, k)-star networks under the comparison diagnosis model. IEEE Trans Reliab 64(1):132–143
42. Yunus NAM, Othman M (2014) Reliability evaluation and routing integration in shuffle exchange omega network. J Netw 9(7):1732–1737
43. Yunus NAM, Othman M (2015) Reliability evaluation for shuffle exchange interconnection network. Procedia Comput Sci 59:162–170
44. Zhu Q, Wang X-K, Cheng G (2013) Reliability evaluation of BC networks. IEEE Trans Comput 62(11):2337–2340

45. Abd-El-Barr M, Gebali F (2014) Reliability analysis and fault tolerance for hypercube multi-computer networks. Inf Sci 276:295–318
46. Rajkumar S, Goyal NK (2014) Design of 4-disjoint gamma interconnection network layouts and reliability analysis of gamma interconnection networks. J Supercomputing 69(1):468–491
47. Sangeetha RG, Chandra V, Chadha D (2014) Bidirectional data vortex optical interconnection network: BER performance by hardware simulation and evaluation of terminal reliability. J Lightwave Technol 32(19):3266–3276
48. Dash RK et al (2012) Network reliability optimization problem of interconnection network under node-edge failure model. Appl Soft Comput 12(8):2322–2328
49. Tripathy PK, Dash RK, Tripathy CR (2015) A dynamic programming approach for layout optimization of interconnection networks. Eng Sci Technol Int J 18(3):374–384
50. Yunus NAM, Othman M (2014) Fault tolerance reliability evaluation in multistage interconnection network. In: Proceedings of the International Conference on Frontiers of Communications, Networks and Applications (ICFCNA), pp 1–5
51. Yunus NAM et al (2016) Reliability review of interconnection networks. IETE Tech Rev, 1–11
52. Yunus NAM, Othman M, Hanapi ZM (2012) Integration of zero and sequential algorithm in shuffle exchange with minus one stage. In: Proceedings of the international conference on Advances in Computing, Control, and Telecommunication Technologies (ACT), pp 7–12
53. Bistouni F, Jahanshahi M (2016) Reliability analysis of multilayer multistage interconnection networks. Telecommun Syst 62(3):529–551
54. Zhou J-X et al (2015) Symmetric property and reliability of balanced hypercube. IEEE Trans Comput 64(3):876–881
55. Dally WJ, Towles BP (2004) Principles and practices of interconnection networks. Morgan Kaufmann, San Francisco
56. Klavžar S, Ma M (2014) Average distance, surface area, and other structural properties of exchanged hypercubes. J Supercomputing 69(1):306–317
57. Rajput IS et al (2012) An efficient parallel searching algorithm on Hypercube Interconnection network. In: 2nd IEEE international conference on Parallel Distributed and Grid Computing (PDGC)
58. Lai C-N (2012) Optimal construction of all shortest node-disjoint paths in hypercubes with applications. IEEE Trans Parallel Distrib Syst 23(6):1129–1134
59. Kuo C-N (2015) Every edge lies on cycles embedding in folded hypercubes with vertex-fault-tolerant. Theoret Comput Sci 589:47–52
60. Abd-El-Barr M, Gebali F (2014) Reliability analysis and fault tolerance for hypercube multi-computer networks. Inf Sci 276:295–318
61. Zhou J-X et al (2015) Symmetric property and reliability of balanced hypercube. IEEE Trans Comput 64(3):876–881
62. Liu Y-L (2015) Routing and wavelength assignment for exchanged hypercubes in linear array optical networks. Inf Process Lett 115(2):203–208
63. Zhang J et al (2015) Dynamic wavelength assignment for realizing hypercube-based Bitonic sorting on wavelength division multiplexing linear arrays. Int J Comput Math 92(2):218–229

Chapter 3
Blocking Problem

3.1 Introduction

In the early 50s, Neumann suggested a simple cost-effective design for electronic computers in which a single processing unit was connected to a single memory module. During the 60s, using the concept of solid-state components, the cost of large computing machines fell. Once it, very large-scale integration (VLSI) is evolved in which thousands of transistors placed on a single chip. Supercomputers were successfully deal with scientific issues such as climate modeling, aerodynamic aircraft design, and particle physics, which created a strong incentive for the development of parallel computers. After the 80s, this technology has played an undeniable role in solving other challenging issues [1].

As discussed in Chap. 1, processors, memory hierarchy, and the interconnection network are vital parts of a parallel system. In other words, the design of an efficient interconnection network is crucial for the efficient construction of multiprocessor systems [2–5].

One of the important factors for choosing a proper interconnection topology is to take into account the blocking problem. If a network is able to handle all possible requests each of which are as a permutation (i.e., defined as a request for parallel connections of each N sources to N corresponding distinct destinations), then the network is non-blocking. As a result, a network is blocking, if it is unable to handle all such requests without conflict and blocking [2, 6–8].

So far, a large number of interconnection topologies have been introduced. However, a few of them can efficiently resolve the blocking problem. For systems with N terminal nodes, a topology would be ideal, if that can connect these nodes by a single switch of size $N \times N$. This type of topology is known as crossbar. In a crossbar network, any processor in the system can connect to any other processor or memory module so that many processors can communicate simultaneously without contention. Clearly, a crossbar network is strictly non-blocking to any permutation of connections. Here, a question arises is that if the crossbar network is strictly

© Springer International Publishing AG, part of Springer Nature 2018   41
M. Jahanshahi and F. Bistouni, *Crossbar-Based Interconnection Networks*,
Computer Communications and Networks,
https://doi.org/10.1007/978-3-319-78473-1_3

non-blocking, whether blocking problem can be considered as a solved problem? Unfortunately, the answer is no. There is an important problem in the use of crossbar network, namely scalability. The number of available pins and the area of the wiring make limit in the size of the largest crossbar implemented by a single chip. Although the technology of VLSI can integrate the crossbar switch hardware into a single chip, the number of pins within a single VLSI chip cannot exceed a certain number [2, 6, 9]. The scalability problem will prevent the direct use of crossbar network for large-size systems. Therefore, crossbar network can be used in small-size multiprocessor systems in practice. To tackle this issue, there is a reasonable solution to take advantage of crossbar networks in systems with large sizes. This solution offers the use of small-size crossbars as building blocks for networks with larger sizes. By studying the pervious works, it can be deduced that this solution can be implemented by two different approaches:

(1) Designing different scalable interconnection networks topology compared to crossbar, using small-size crossbar networks as switching elements [2, 6, 7, 10–32]. So far, a large number of interconnection topologies are designed using this approach that most of them are known as multistage interconnection networks. (2) Designing scalable crossbar networks using small-size crossbar networks as switching elements [2, 6, 33]. This approach can lead to the design of scalable networks, which are non-blocking similar to crossbar network.

In the remainder of this chapter, the two aforementioned approaches will be discussed more detailed. Next, in Chap. 4 on behalf of the first approach, several approaches will be introduced to improve fault-tolerance metric (as a way to improve blocking problem) in multistage interconnection networks. Then, in Chap. 5, a new non-blocking interconnection topology will be proposed on behalf of the second approach.

3.2 Related Works

As discussed in the previous section, crossbar network suffers from scalability problem to exploit in large-size systems. The main reason for this problem is that a large number of pins are required to implement a large-size crossbar network on a single VLSI chip. However, the number of pins in a VLSI chip cannot exceed a few hundred. This will result in restrictions on the size of crossbar network. The solution to deal with this problem is the use of small-size crossbars as building blocks of larger networks. On the other hand, two different scenarios can implement this solution: (1) Take advantage of small-size crossbar networks as switching elements to build larger networks that differ with crossbar, topologically [2, 6, 7, 10–32]. (2) Take advantage of small-size crossbar networks as switching elements to build larger crossbar networks that are equivalent to the crossbar, topologically [2, 6, 33]. We will discuss these two different scenarios in sub-Sects. 3.2.1 and 3.2.2, respectively.

3.2.1 Construction of Scalable Non-crossbar Networks by Small-Size Crossbars

This approach can be used as a base for making different topologies. In what follows, important interconnection topologies made based on this approach have been investigated:

Generally, when it is discussed about the banyan-type network in the literature, the purpose of it is a typical multistage interconnection network (MIN) that can provide only a single path between each pair of source–destination. Therefore, the network faces with single-point-of-failure and fails in the event of a failure in one of its components. So far, a variety of banyan-type topologies such as shuffle-exchange network, omega network, baseline network, binary n-cube network, and delta networks are presented by researchers. The remarkable thing is that all of these types of networks typically have used 2×2 crossbar networks as their switching elements.

A MIN called gamma has been presented in [12]. A gamma network of size 8×8 is shown in Fig. 3.1. This network can establish connection between N source nodes to N destination nodes. This network made up of $(\log_2 N + 1)$ stages, which are numbered from 0 through $(\log_2 N)$. In addition, it uses small-size crossbar networks as switching elements on each stage. The number of crossbar switches used on each stage is equal to N. Also, the crossbar switches used in the first, last, and middle stages are of small size 1×3, 3×1, and 3×3, respectively. Gamma network can provide different paths for many source–destination pairs, as this can help to fault-tolerance capability. However, it cannot provide more than one path, when the tag number is same for the source and destination. In these cases, the network will be the single-point-of-failure, which is non-fault-tolerant. Therefore, gamma has different levels of reliability to different terminal nodes and cannot guarantee fault-tolerance capability for all scenarios.

In [13], two new designs of the 4-disjoint paths MINs namely 4-disjoint gamma interconnection networks (4DGIN-1 and 4DGIN-2) have been proposed in order to improve fault tolerance and reliability of gamma network. The 4DGIN-1 and 4DGIN-2 networks are shown in Figs. 3.2 and 3.3, respectively. Consider two 4DGIN-1 and 4DGIN-2 network of size $N \times N$. The number of switching stages in these new topologies is equal to $(\log_2 N + 1)$, which are numbered from 0 to $(\log_2 N)$. Also, the small-size crossbar networks have been used in any of the switching stages of these fault-tolerant networks.

A new fault-tolerant MIN topology is proposed in [14] called Combining Switches Multistage Interconnection Network (CSMIN). To meet the fault-tolerant metric, there are two different paths between each pair of source–destination. When one of the paths fails, then the other path dynamically can be used as a successor route for forwarding packets to improve the blocking situation. A CSMIN of size 8×8 is shown in Fig. 3.4. Consider a CSMIN network of general size $N \times N$. This topology has $(\log_2 N + 1)$ stages, which are numbered from 0 to $(\log_2 N)$. CSMIN network uses small-size crossbar networks as the switching elements in

Fig. 3.1 A gamma network
of size 8 × 8

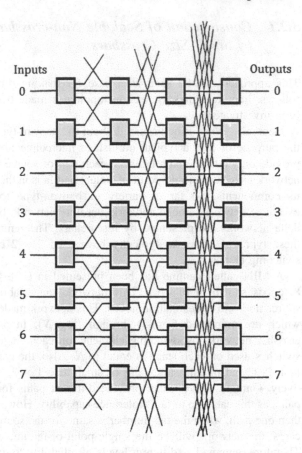

each of the stages. Size of the crossbar switches in the first and last stage is 2 × 4 and 3 × 2, respectively. In addition, size of them in the stage 1 and intermediate stage is 3 × 3 and 4 × 4, respectively.

In [15], a new topology called Fault-tolerant Fully-Chained Combining Switches Multistage Interconnection Network (FCSMIN) has been introduced to eliminate the backtracking penalties of the CSMIN. Figure 3.5 shows a FCSMIN of size 8 × 8. Since FCSMIN provides several different paths between each pair of source–destination, it is able to meet the fault-tolerance parameter. In the FCSMIN, in stages 1 to $(\log_2 N - 1)$, one of the original non-straight links of CSMIN has been changed to a chained link. In addition, either of the non-straight links between the last two stages of CSMIN has been removed in FCSMIN structure. Generally, a FCSMIN of size $N \times N$ has $(\log_2 N + 1)$ switching stages that are numbered from 0 to $(\log_2 N)$. This network also takes advantage of the small-size crossbars within its structure. Size of the crossbar switches in the first stage, the middle stages, and the last stage is 2 × 4, 3 × 3, and 2 × 1, respectively.

Wei and Lee [16] introduces a new MIN topology that can provide fault-tolerance and reliability parameters. An 8 × 8 EGN is shown in Fig. 3.6. Let us

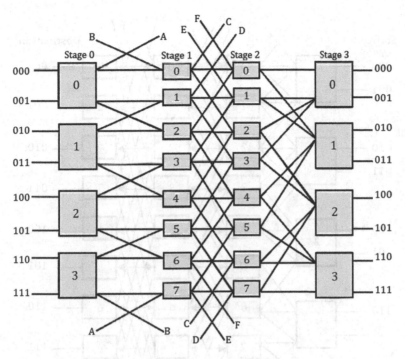

Fig. 3.2 A 4DGIN-1 network of size 8 × 8

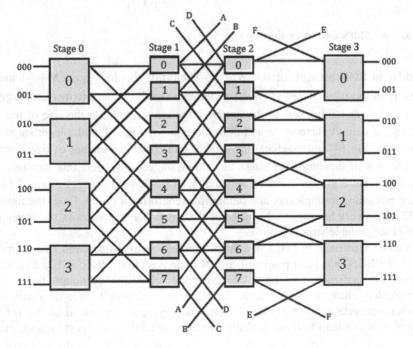

Fig. 3.3 A 4DGIN-2 network of size 8 × 8

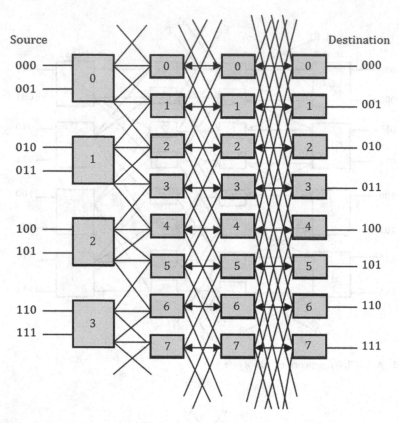

Fig. 3.4 A CSMIN network of size 8×8

consider an EGN network of size $N \times N$. Then, this topology has $\left(N + \frac{N}{m}\right)$ multiplexers in the input stage, $\left(\frac{N}{2} + \frac{N}{2m}\right)$ switches in each of the intermediate stages, and $\left(N + \frac{N}{m}\right)$ demultiplexers in the output stage (m is related to the size of multiplexers and demultiplexers). In addition, there are one $m \times 1$ multiplexer stage as input stage, $\log_2 \left(\frac{N}{m}\right)$ intermediate stages including crossbar switches of size 2×2, and one $1 \times m$ demultiplexer stage as output stage. Moreover, this structure is classified in some groups in such a way that each single path network of size $\frac{N}{m}$ plus its corresponding multiplexers and demultiplexers forms a group. From the figure, the EGN network has taken advantage of the small-size crossbars in the development of its scalable structure.

In another research in [7], a new MIN topology named Improved Extra Group Network (IEGN) has been proposed. IEGN is derived from EGN and its aim is to improve the parameters of fault tolerance and reliability. Although the IEGN structure has changed compared to EGN, it still adheres to use of small-size crossbar networks in its topology. Some auxiliary links are added to the IEGN network that can help to improve fault tolerance, reliability, and performance even

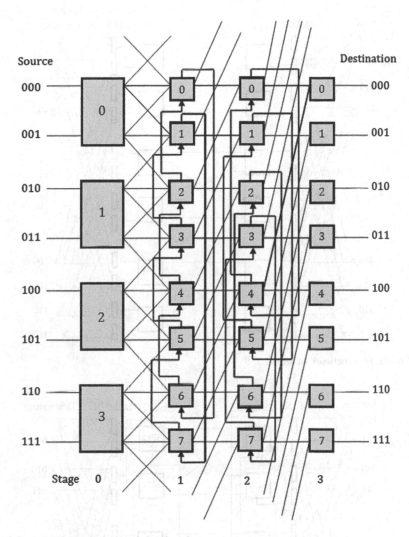

Fig. 3.5 A FCSMIN network of size 8×8

in the presence of faults. The auxiliary links have caused the size of the crossbars is changed from 2×2 to 3×3. An IEGN of size 8×8 is shown in Fig. 3.7. An IEGN of size $N \times N$ has $\left(N + \frac{N}{m}\right)$ multiplexers in the input stage, $\left(\frac{N}{2} + \frac{N}{2m}\right)$ crossbar switches in each of the middle stages, and $\left(N + \frac{N}{m}\right)$ demultiplexers in the output stage (Here, m is related to the size of multiplexers and demultiplexers.). In addition, IEGN uses one $m \times 1$ multiplexer stage as input stage, $\log_2\left(\frac{N}{m}\right)$ intermediate stages of 3×3 crossbar switches, and one $1 \times m$ demultiplexer stage as output stage.

A new topology of interconnection networks called Hierarchical Adaptive Switching Interconnection Network (HASIN) has been introduced in [17].

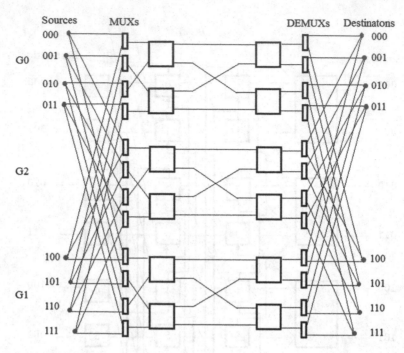

Fig. 3.6 An EGN network of size 8×8

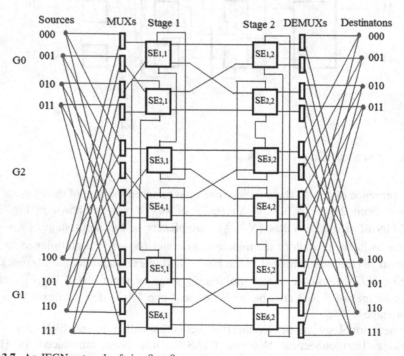

Fig. 3.7 An IEGN network of size 8×8

In general, this topology has two structural levels namely local level and global level. Local level makes use of small-size crossbar networks and general level uses the mesh network. Figure 3.8 shows an example of HASIN topology consisting of 28 cores. This hierarchical structure reduces the number of hops and explores the communication locality. In addition, since small-size crossbar switches have an efficient structure with no need to buffer, the power consumption is much less compared to conventional router structure. Therefore, the use of small-size crossbar networks is a good idea to improve the performance in HASIN.

A new fault-tolerant MIN called Augmented Shuffle-Exchange Network (ASEN) was proposed in [18]. The main objective of this research was to improve the reliability and fault tolerance compared to banyan-type networks such as shuffle-exchange network (SEN). ASEN of size 8 × 8 is shown in Fig. 3.9. As can

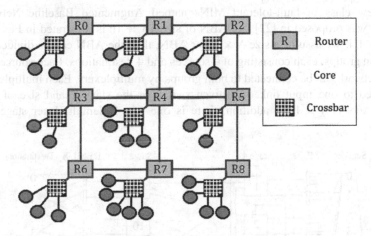

Fig. 3.8 HASIN topology consisting of 28 cores

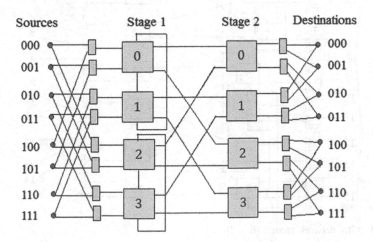

Fig. 3.9 ASEN network of size 8 × 8

be seen, the ASEN is made up of small-size crossbar networks as switching elements. In fact, the ASEN is a SEN but one of the switching stages has been removed. In addition, some auxiliary links, multiplexers, and demultiplexers have been added to the ASEN. Suppose that size of ASEN is $N \times N$. In this general case, ASEN will have $((\log_2 N) - 1)$ stages so that each stage includes $\left(\frac{N}{2}\right)$ crossbar switches. The size of the crossbar switches used in stages 1 through $((\log_2 N) - 2)$ and last stage is 3×3 and 2×2, respectively. In addition, there is one 2×1 multiplexer stage as input stage before switching stage 1 and one 1×2 demultiplexer stage as output stage after switching stage $((\log_2 N) - 1)$. The number of multiplexers and demultiplexers is equal to N for ASEN of size $N \times N$. Let us define network complexity as the number of 2×2 switching elements in the network. As a result, the network complexity of an $N \times N$ ASEN is equal to $\left[\left(\frac{3N}{2}\right)\left(1 + \frac{3}{4}((\log_2 N) - 2)\right)\right]$.

A new class of fault-tolerant MINs named Augmented Baseline Networks (ABNs) was proposed in [22]. An ABN of size 16×16 is illustrated in Fig. 3.10. In general, if we assume a size $N \times N$ for ABN, then the ABN can be divided into two main groups, each consisting of $\frac{N}{2}$ sources and $\frac{N}{2}$ destinations. Each source node in the network can be connected to both groups by multiplexers. Each multiplexer is connected to one input link of a given switch in the stage 1 and size of these multiplexers is 4×1. In addition, there is one 1×2 demultiplexer stage after

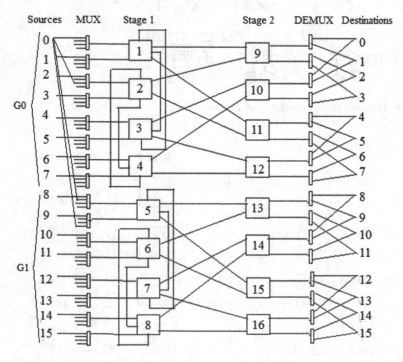

Fig. 3.10 ABN network of size 16×16

stage $((\log_2 N) - 2)$ so that each output link of a switch in stage $((\log_2 N) - 2)$ is connected to one demultiplexer. It should be noted that ABN uses small-size crossbar network as switching elements in its switching stages. Size of the switches used in stages 1 through $((\log_2 N) - 3)$ is 3×3, and the size of the switches used in last switching stage is 2×2. The network complexity of an $N \times N$ ABN is given by $\left[\left(\frac{9N}{8}\right)\left(\frac{16}{9}((\log_2 N) - 3)\right) \right]$.

Another important class of interconnection networks that are considered by many researchers in this area is replicated MINs [7, 27]. Generally, replicated MIN refers to a network that is derived from a banyan-type MIN so that the banyan-type MIN is replicated in L layers. Therefore, the replicated MINs can provide several different paths between any source–destination pairs that offers hope for a better fault tolerance and reliability compared to the banyan-type MINs. Figure 3.11 shows the architecture of such an 8×8 replicated MIN consisting of two layers $(L = 2)$ in a three-dimensional view. Consider L-layer replicated MIN of size $N \times N$. In this general case, number of stages in the network will be equal to $(\log_2 N)$. In addition, all switches used in these stages are small-size crossbars, mainly 2×2 crossbars. It is worth noting that the number of crossbar switches in the network according to the number of network layers is equal to $\left(\frac{L \times N}{2}\right)$. In this topology, there is one $1 \times L$ demultiplexer for each L input links of peer switches located in stage 1 and one $L \times 1$ multiplexer for each L output links of peer switches located in stage $(\log_2 N)$. Therefore, the number of demultiplexers and multiplexers is equal to N. The network complexity for an $N \times N$ L-layer replicated MIN is calculated as $\left[\left(\frac{L \times N}{2}\right)(1 + (\log_2 N)) \right]$.

As can be seen, all topologies discussed in this subsection use the small-size crossbar networks as their constituent elements. Since each of these small-size crossbar switches can be implemented in a single chip, scalability problem can be easily solved. However, these topologies that are mostly of the type of multistage interconnection networks are incapable of providing an efficient solution to the blocking problem. Below, we will examine the reasons of this issue:

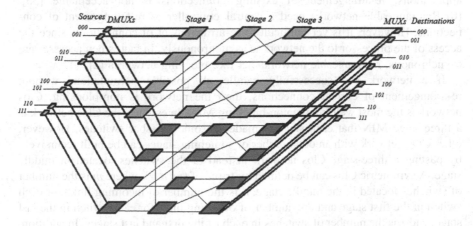

Fig. 3.11 Three-dimensional perspective of an 8×8 two-layer replicated MIN

In a general view, MINs can be divided into two main groups: (1) single-path MINs (banyan-type) and (2) multipath MINs. Single-path MINs are the ones that cannot provide more than one path between each pair of source–destination. This structure can lead to the blocking problem, since the request for a new connection may be impossible due to busy resources such as links and switches by other existing connections. That is why these networks are known as blocking MINs. In contrast to these single-path networks, there are multipath MINs that can provide more than one path between each pair of source–destination. When a path is not available, then the network can switch to alternative path to handle a connection request. As a result, the existence of multiple paths in the network can improve fault-tolerance capability and reduce the occurrence of the blocking problem. According to this argument, one of the approaches to alleviate the blocking problem is improving fault-tolerance feature on the network. For this reason, fault tolerance in MINs is one of the favorite topics among researchers. Therefore, because of the importance of fault-tolerance parameter in multistage interconnection networks, the next chapter is devoted to a discussion on some new approaches to improve this feature in these networks.

Fault-tolerant MINs are of interest because of cost-effectiveness compared to crossbar network. However, most of these networks cannot provide a fully non-blocking mode that needs to manage all conflicts. Thus, most of this kind of MINs are also considered as blocking MINs. Only two classes of fault-tolerant MINs that may be able to solve the blocking problem are as follows: (1) rearrangeable non-blocking (or simply rearrangeable) MINs such as Benes network [31] and $(2n - 1)$-Stage Shuffle-Exchange Networks $(n = \log_2 N)$ [28–30] and (2) non-blocking Clos network [32].

The main idea of the rearrangeable network to deal with the blocking problem is re-arrangement of connections. This network can respond to all connections requests in every permutation by re-arrangement of current connections if needed. At first glance, these networks seem to be efficient solution in theory. Nevertheless, there are some problems in practice with a closer look: (a) In uninterruptible applications, re-arrangement of existing connections is not acceptable [33]. (b) Rearrangeable networks need a central controller to re-arrangement of connections. However, it is very difficult to re-arrangement of connections, since the access of the processor to the network is asynchronously. In fact, when accesses are asynchronous, rearrangeable networks act like blocking networks [2].

If a network can successfully handle all possible permutations without re-arrangement of current connections, then the network is non-blocking. Clos network is the most well-known non-blocking MIN. In essence, the Clos network is a three-stage MIN that each stage is made of some crossbar switches. However, other Clos network with an odd number of switching stages can be built recursively by pasting a three-stage Clos network instead of the switches located in middle stage. A symmetric Clos can be defined by triple of (m, n, r), where m is the number of switches located in the middle stage, n is the number of incoming links for each switch in the first stage and the number of outgoing links for each switch in the last stage, and r is the number of switches in each of the first and last stages. In addition, size of crossbar switches located on the first stage, the middle stage, and the last stage is $n \times m$, $r \times r$, and $m \times n$, respectively. Although the Clos network can be non-blocking in theory, there are some important issues in the way of this solution in practice: (a) It has been proven that Clos is non-blocking if this condition be true: $m \geq 2n - 1$ [6, 16]. Therefore, there are some structural constraints for a non-blocking Clos. (b) An efficient control mechanism for the allocation of connections in the Clos is essential. However, this mechanism is usually complex in a Clos network [6, 19, 33–36]. For routing a packet in the Clos, after it was sent to switch on the first stage, each switch in the middle stage can be considered for forwarding packet, as long as the link connected to the switch is free. Also, this middle switch should choose a free link to switch on the last stage. Here, when this link is busy, the path is impossible. Finally, switch on the last stage should choose the selected outgoing link. Thus, the problem of routing in the Clos is largely dependent on an efficient mechanism for the allocation of each packet to a middle-stage switch.

Altogether, according to the discussions in this section, it can be said that MINs can solve the scalability problem raised in the crossbar network. On the other hand, although these networks cannot fully cope with the blocking problem, they can support the important metric of fault tolerance that can result in reducing the blocking problem. Therefore, fault-tolerant MINs are of particular importance in this area. In Chap. 4, some important methods to improve the fault-tolerance metric on MINs will be examined.

3.2.2   Construction of Scalable Crossbar Topologies by Small-Size Crossbars

In this approach, the idea is to build large-size crossbar networks using the small-size networks as switching elements. This approach can have several important advantages: First, the blocking problem can easily be solved, because the crossbar networks are strictly non-blocking. Second, the issue of scalability can be solved by the use of small-size crossbars. Therefore, this approach can be a more efficient solution compared to that approach discussed in the previous subsection. In this area, some ideas can be found in [2, 6]. However, the number of topologies designed based on this approach is very low. A rare instance of this type of topology is Multistage Crossbar Network (MCN) [33]. MCN is a multistage implementation of crossbar architecture and it uses small-size crossbar networks as switching elements.

Figure 3.12 shows a MCN of size 4×4, which made up of some 2×2 crossbar switches. In a general case, consider a MCN of size $N \times N$. The structure is composed of (N^2) crossbar switches of size 2×2. Such crossbar-based networks are useful to build large crossbar networks, promoting scalability due to exploitation of small-size crossbar switches as switching elements. Therefore, thanks to using the small-size crossbars, MCN can solve the problem of scalability. For some source–

Fig. 3.12 A MCN of size
4 × 4

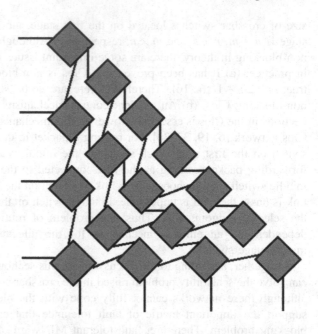

destination pairs in MCN, the number of paths is more than one. The path length is defined as the number of switching elements between a source–destination pair. In the MCN, path length is not the same for all origin–destination pairs. More precisely, the path length could be a number between 1 through $(2N - 1)$ in the MCN.

MCN can be a reasonable solution to the problem of scalability. However, disadvantage of this structure is its high hardware cost. The hardware cost of a network can be calculated based on the total number of crosspoints in it [7, 10, 22, 27, 35]. According to this definition, the hardware cost of the crossbar network is equal to N^2. In addition, the number of 2×2 crossbar switches in the MCN is equal to N^2; thus, its cost is $4N^2$. The cost of the MCN is not acceptable, because it is four times higher than the cost of typical crossbar network.

In view of the discussions that took place in this subsection, design of scalable crossbar networks using small-size crossbar switches is a good idea for construction of scalable non-blocking interconnection networks. However, in a few interconnection topologies, this approach has been used in their design. MCN is a network built based on this approach and it can solve the problems of blocking and scalability. However, it has four times higher hardware cost compared to typical crossbar network. Therefore, other designs should be provided to achieve better performance compared to MCN in terms of cost. For this purpose, a new interconnection topology named Scalable Crossbar Network (SCN) will be discussed in Chap. 5. SCN is designed based on the approach discussed in this subsection and it brings the following advantages: (1) It is a non-blocking network. (2) It can solve the problem of scalability using small-size crossbar networks as switching elements in its structure. (3) It has a same hardware cost compared to typical crossbar network.

As will be discussed in Chap. 5, SCN can meet all the above requirements. In addition, the routing mechanism for the SCN network can be fast, cost-effectiveness, and self-routing. In addition, performance analysis conducted in Chap. 5 demonstrates that the SCN can obtain very good performance in terms of various important metrics including terminal reliability, mean time to failure, and system failure rate compared to different interconnection topologies namely SEN, SEN+, Benes network, replicated MIN, multilayer MINs, and MCN.

References

1. Jadhav SS (2009) *Advanced computer architecture and computing*. Technical Publications
2. Duato J, Yalamanchili S, Ni LM (2003) Interconnection networks: an engineering approach. Morgan Kaufmann, USA
3. Dubois M, Annavaram M, Stenström P (2012) *Parallel computer organization and design*. Cambridge University Press, Cambridge
4. Culler DE, Singh JP, Gupta A (199) *Parallel computer architecture: a hardware/software approach*. Morgan Kaufmann
5. Agrawal DP (1983) Graph theoretical analysis and design of multistage interconnection networks. IEEE Trans Comput 100(7):637–648
6. Dally WJ, Towels BP (2004) Principles and practices of interconnection networks. Morgan Kaufmann, San Francisco, Calif, USA
7. Bistouni F, Jahanshahi M (2014) Improved extra group network: a new fault-tolerant multistage interconnection network. J Supercomput 69(1):161–199
8. Villar JA et al (2013) An integrated solution for QoS provision and congestion management in high-performance interconnection networks using deterministic source-based routing. J Supercomput 66(1):284–304
9. Hur JY et al (2007) Systematic customization of on-chip crossbar interconnects. *Reconfigurable computing: architectures, tools and applications*. Springer Berlin Heidelberg, pp 61–72
10. Bistouni F, Jahanshahi M (2015) Pars network: a multistage interconnection network with fault-tolerance capability. J Parallel Distrib Comput 75:168–183
11. Bistouni F, Jahanshahi M (2014) Analyzing the reliability of shuffle-exchange networks using reliability block diagrams. Reliab Eng Syst Saf 132:97–106
12. Parker DS, Raghavendra CS (1984) The gamma network. IEEE Trans Comput 100(4): 367–373
13. Rajkumar S, Goyal NK (2014) Design of 4-disjoint gamma interconnection network layouts and reliability analysis of gamma interconnection networks. J Supercomput 69(1):468–491
14. Chen CW, Chung CP (2005) Designing a disjoint paths interconnection network with fault tolerance and collision solving. J Supercomput 34(1):63–80
15. Nitin SG, Srivastava N (2011) Designing a fault-tolerant fully-chained combining switches multi-stage interconnection network with disjoint paths. J Supercomput 55(3):400–431
16. Wei S, Lee G (1988) Extra group network: a cost-effective fault-tolerant multistage interconnection network. ACM SIGARCH Comput Archit News 16(2) IEEE Computer Society Press
17. Matos D et al (2013) Hierarchical and multiple switching NoC with floorplan based adaptability. Reconfigurable computing: architectures, tools and applications. Springer, Berlin, Heidelberg, pp 179–184
18. Kumar VP, Reddy SM (1987) Augmented shuffle-exchange multistage interconnection networks. Computer 20(6):30–40

19. Vasiliadis DC, Rizos GE, Vassilakis C (2013) Modelling and performance study of finite-buffered blocking multistage interconnection networks supporting natively 2-class priority routing traffic. J Netw Comput Appl 36(2):723–737
20. Gunawan I (2008) Reliability analysis of shuffle-exchange network systems. Reliab Eng Syst Saf 93(2):271–276
21. Blake JT, Trivedi KS (1989) Reliability analysis of interconnection networks using hierarchical composition. IEEE Trans Reliab 38(1):111–120
22. Bansal PK, Joshi RC, Singh K (1994) On a fault-tolerant multistage interconnection network. Comput Electr Eng 20(4):335–345
23. Blake JT, Trivedi KS (1989) Multistage interconnection network reliability. IEEE Trans Comput 38(11):1600–1604
24. Nitin, Subramanian A (2008) Efficient algorithms and methods to solve dynamic MINs stability problem using stable matching with complete ties. J Discrete Algorithms 6(3):353–380
25. Fan CC, Bruck J (2000) Tolerating multiple faults in multistage interconnection networks with minimal extra stages. IEEE Trans Comput 49(9):998–1004
26. Adams GB, Siegel HJ (1982) The extra stage cube: a fault-tolerant interconnection network for supersystems. IEEE Transac Comput 100(5):443–454
27. Tutsch D, Hommel G (2008) MLMIN: a multicore processor and parallel computer network topology for multicast. Comput Oper Res 35(12):3807–3821
28. Çam H (2001) Analysis of shuffle-exchange networks under permutation trafic. Switching networks: recent advances. Springer, USA, pp 215–256
29. Çam H (2003) Rearrangeability of (2n − 1)-stage shuffle-exchange networks. SIAM J Comput 32(3):557–585
30. Dai H, Shen X (2008) Rearrangeability of 7-stage 16 × 16 shuffle exchange networks. Front Electr Electron Eng China 3(4):440–458
31. Beneš VE (1965) Mathematical theory of connecting networks and telephone traffic, vol 17. Academic Press
32. Clos C (1953) A study of non-blocking switching networks. Bell Syst Tech J 32(2):406–424
33. Kolias C, Tomkos I (2005) Switch fabrics. IEEE Circ Devices Mag 21(5):12–17
34. Fey D et al (2012) Optical multiplexing techniques for photonic Clos networks in high performance computing architectures. J Supercomput 62(2):620–632
35. Cuda D, Giaccone P, Montalto M (2012) Design and control of next generation distribution frames. Comput Netw 56(13):3110–3122
36. Sibai FN (2011) Design and evaluation of low latency interconnection networks for real-time many-core embedded systems. Comput Electr Eng 37(6):958–972

Chapter 4
Fault-Tolerant Multistage Interconnection Networks

4.1 Introduction

As mentioned in the previous chapter, the scalability problem in crossbar networks can be solved by using multistage interconnection networks (MINs). On the other hand, presenting novel network topologies that have been improved in terms of the fault-tolerant capability improves the blocking problem to a satisfactory level. As a result, one of the most interesting topics in this field of research is promoting the fault tolerance in these networks.

Developing methods that could increase the number of paths from any source node to any destination node is the main idea to improve the fault tolerance of interconnection networks. If there is unavailability of primary paths due to being busy or failure of one or more switches along the path, using alternative routes would be possible to increase the number of paths across the networks. In consequence, it reduces the blocking issue in the network. Furthermore, other performance parameters including throughput can be improved by this method. Generally, fault tolerance can enable the network to maintain its performance, regardless of the availability of some of the switching elements or network links.

The rest of this chapter is formed in this way: The next section (Sect. 4.2) will discuss about the reliability parameters in one of the most important methods to improve fault tolerance of interconnection networks. The number of switching stages is increased in this method, which can increase the number of paths from any source node to any destination node in the network. Then, in Sect. 4.3, a new topology of the fault-tolerant MINs will be introduced that it is affordable as well; it has a good performance regarding different performance parameters. Subsequently, Sect. 4.4 will discuss the reliability improving in Benes network as one of the most important MINs. Also, it will introduce how to improve its reliability in the best way.

© Springer International Publishing AG, part of Springer Nature 2018   57
M. Jahanshahi and F. Bistouni, *Crossbar-Based Interconnection Networks*,
Computer Communications and Networks,
https://doi.org/10.1007/978-3-319-78473-1_4

4.2 Performance Analysis of Increasing the Number of Stages in Terms of Reliability

Creation of redundant paths from any source node to any destination node is considered as the main approach to improve the fault tolerance of MINs. Moreover, one of the main ideas is that the increased number of stages leads to create redundancy in MINs' paths [1–3]. However, the effectiveness of fault-tolerant systems can be proved by reliability as one of the basic parameters [4]. In other words, improvement of reliability could be the result of an efficient method for improving fault tolerance. Here the question is what is the impact of increasing the number of stages on reliability? The positive impact on the reliability of MINs by adding one extra stage has been shown in the previous studies [1, 5, 6]. However, the impact of increasing one more stage on the obtained reliability is another disputation here. To answer these questions, investigation of the reliability of three MINs, Shuffle-Exchange Network (SEN), SEN with one extra stage (SEN+), and SEN with two extra stages (SEN+2), in a recent study has been presented in the following conclusions [7]:

(1) SEN+ always has a higher reliability than SEN and SEN+2.
(2) SEN achieves much higher reliability than SEN+2.

According to the high network complexity of the SEN+2, the first result is quite expected. But since the previous studies have been introduced the single-path MINs as the most unreliable networks, and considering that these networks cannot be efficient in case of a single fault, the second result is surprising [1–3, 5–9]. In addition to this unexpected conclusion [7], there are also some other issues:

(1) Most of reliability equations are computed for a small network size 8 × 8.
(2) To obtain a correct result, there is usually need to use some tools like reliability block diagrams due to the presence of complex relationships between network components [6, 9]. However, the analysis carried out in [7] is limited to a simple description of the relationships that cannot be used to determine the reliability of the MIN [7].

In summary, based on above discussions, the aim of our study is to re-analyze the reliability of three aforementioned networks with more information. The evaluation is performed in this research will represent single-path SEN as one of the most unreliable MINs.

4.2.1 Structure of the SEN, SEN+, and SEN+2

The structure of Shuffle-Exchange Network will be discussed in this subsection. SENs of size $N \times N$ are comprised of $(\log_2 N)$ stages, in which each stage contains $\left(\frac{N}{2}\right)$ switching elements of size 2×2. In Fig. 4.1, a SEN of size 8×8 is represented. The network complexity of an $N \times N$ SEN is equal to $\left(\frac{N}{2}(\log_2 N)\right)$. Only one path can be provided between each source–destination pair in this network.

A SEN with an extra stage is called as SEN+. Generally, a SEN+ of size $N \times N$ has $((\log_2 N) + 1)$ stages, while each stage is made up of $\left(\frac{N}{2}\right)$ crossbar switches of size 2×2. Figure 4.2 represents SEN+ of size 8×8. This network can provide two different paths between each pair of source–destination. In addition, the network complexity of a SEN+ of size $N \times N$ is equal to $\left(\frac{N}{2}\right) \times ((\log_2 N) + 1)$.

A SEN with two additional stages is called as SEN+2. Generally, a SEN+2 of size $N \times N$ is comprised of $((\log_2 N) + 2)$ stages, while each stage contains $\left(\frac{N}{2}\right)$ switching elements of size 2×2. Figure 4.3 represents a SEN+2 of size 8×8. The network complexity of an $N \times N$ SEN+2 network is $\frac{N}{2}((\log_2 N) + 2)$. Four different paths between each pair of source–destination can be provided by this network.

4.2.2   Terminal Reliability of SEN, SEN+, and SEN+2

Terminal reliability refers to the probability of at least one fault-free path between a source–destination pair. Therefore, it represents the reliability between separate pairs of sources and destinations.

All switches between a source–destination pair are essential, as SEN is a single-path MIN. Figure 4.4 represents the terminal reliability RBD (reliability block diagram) of SEN of size $N \times N$ is.

Let r be the probability of a switch being operational. Terminal reliability of $N \times N$ SEN is obtained by Eq. (4.1).

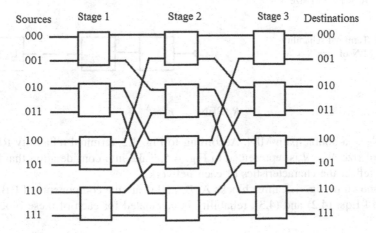

Fig. 4.1 A SEN of size 8×8

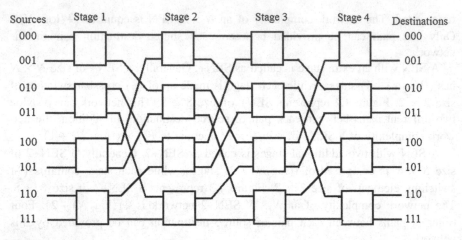

Fig. 4.2 A SEN+ of size 8×8

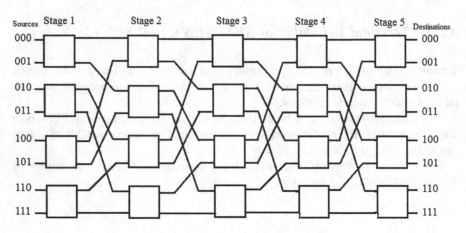

Fig. 4.3 A SEN+2 of size 8×8

Fig. 4.4 Terminal reliability RBD of SEN of size $N \times N$

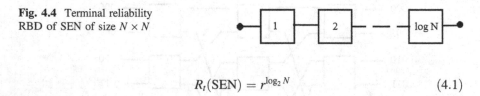

$$R_t(\text{SEN}) = r^{\log_2 N} \tag{4.1}$$

SEN+ is a double-path MIN. According to Fig. 4.2, terminal reliability RBD of SEN+ of size $N \times N$ is represented in Fig. 4.5. Take into consideration that RBDs should reflect the characteristics of each network.

As shown in Fig. 4.5, three blocks A, B, and C are the components of RBD. By means of Eqs. (4.2) and (4.3), reliability is calculated for each of these blocks.

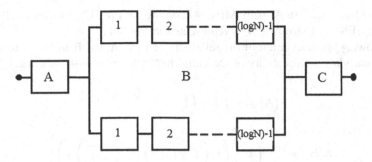

Fig. 4.5 Terminal reliability RBD of SEN+ of size $N \times N$

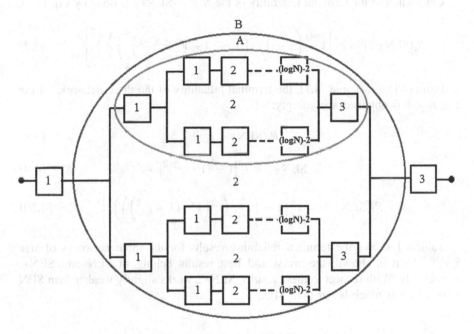

Fig. 4.6 Terminal reliability RBD of SEN+2 of size $N \times N$

$$R(A) = R(B) = r \tag{4.2}$$

$$R(C) = 1 - \left(1 - r^{(\log_2 N) - 1}\right)^2 \tag{4.3}$$

The calculation of terminal reliability of the $N \times N$ SEN+ is as follows:

$$R_t(\text{SEN}+) = r^2 \left[1 - (1 - r^{(\log_2 N) - 1})^2\right] \tag{4.4}$$

SEN+2 is a quadruple-path MIN. According to Fig. 4.3, terminal reliability RBD of SEN+2 of size $N \times N$ is represented in Fig. 4.6.

Following the calculation of the reliability of parts A and B in Fig. 4.6, we are able to calculate the reliability of the entire network.

$$R(\text{A}) = r^2 \left(1 - \left(1 - r^{(\log_2 N)-2} \right)^2 \right) \tag{4.5}$$

$$R(\text{B}) = 1 - \left(1 - \left(r^2 \left(1 - \left(1 - r^{(\log_2 N)-2} \right)^2 \right) \right) \right)^2 \tag{4.6}$$

Calculation of the terminal reliability of the $N \times N$ SEN+2 is done by Eq. (4.7).

$$R_t(\text{SEN+2}) = r^2 \left[1 - \left(1 - \left(r^2 \left(1 - \left(1 - r^{(\log_2 N)-2} \right)^2 \right) \right) \right)^2 \right] \tag{4.7}$$

From (4.1), (4.4), and (4.7), the terminal reliability of the three networks of the size 8×8 is obtained accordingly:

$$R_t(\text{SEN}) = r^3 \tag{4.8}$$

$$R_t(\text{SEN+}) = r^2[1 - (1 - r^2)^2] \tag{4.9}$$

$$R_t(\text{SEN+2}) = r^2 \left[1 - \left(1 - \left(r^2 \left(1 - (1 - r)^2 \right) \right) \right)^2 \right] \tag{4.10}$$

Table 4.1 shows the terminal reliability results for the three networks of size 8×8. As it is shown, the worst and best results belong to SEN and SEN+, respectively. With respect to these results, SEN+2 works slightly weaker than SEN +, though it is much better than SEN.

Table 4.1 Comparison of the terminal reliabilities of 8×8 SEN, SEN+, and SEN+2	Switching reliability	$R_t(\text{SEN})$	$R_t(\text{SEN+})$	$R_t(\text{SEN+2})$
	0.99	0.970299	0.979712	0.979708
	0.98	0.941192	0.958894	0.958865
	0.97	0.912673	0.937614	0.937519
	0.96	0.884736	0.915935	0.915721
	0.95	0.857375	0.893921	0.893519
	0.94	0.830584	0.871628	0.870965
	0.93	0.804357	0.849114	0.848108
	0.92	0.778688	0.826431	0.824997
	0.91	0.753571	0.803630	0.801683
	0.90	0.729000	0.780759	0.778213

We got the same results if we compare our results with the previous studies [7] for SEN and SEN+. But in the previous investigations [7], the results for SEN+2 were unexpectedly imprecise. The method of reliability analysis could be the main reason for this. As discussed in Sect. 4.2, determination of accurate relationships between the components of fault-tolerant MINs is intractable [6]. Furthermore, the simplified description of the relationships that the result gives us is not desirable to identify the reliability of the MIN in general [7]. SEN and SEN+ have less complexity as compared to the structure of the SEN+2; therefore, the simplified model may be suitable to analyze the reliability of such simpler systems. However, using the simple approach for the more complex network leads to increase false results. For this reason, we need a more suitable approach such as the RBD method, to minimize the probability of error.

The results are shown in Fig. 4.7 to provide better comparison. According to terminal reliability results, represented SEN+ and SEN+2 are very similar and additionally they are both more reliable than SEN. One extra stage has more efficiency than adding two stages to SEN regarding to terminal reliability. It is because there is almost no change by adding two stages in terminal reliability of SEN, while it is increased by adding one extra stage.

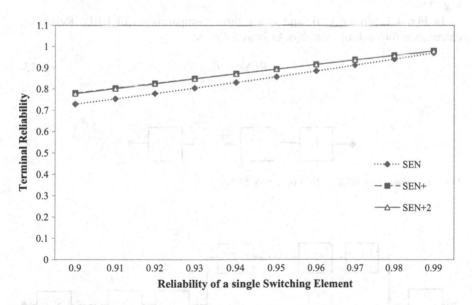

Fig. 4.7 Terminal reliability versus reliability of a single switching element

4.2.3 Broadcast Reliability of SEN, SEN+, and SEN+2

Broadcast reliability is defined as the probability of successful communication between the source and all network destinations, which represents the reliability between a given source and all destinations.

So, as discussed earlier about the network structure of SEN, in this type of reliability all switches in the last stage are essential. Based on Fig. 4.1, broadcast reliability RBD of $N \times N$ SEN is represented in Fig. 4.8.

K in Fig. 4.8 is calculated using Eq. (4.11).

$$K = \sum_{i=1}^{\log_2 N} \frac{N}{2^i} \tag{4.11}$$

Broadcast reliability of $N \times N$ SEN is obtained by Eq. (4.12).

$$R_b(\text{SEN}) = r^K \tag{4.12}$$

Based on Fig. 4.2, broadcast reliability RBD of $N \times N$ SEN+ is represented in Fig. 4.9.

In Fig. 4.9, blocks A, B, and C are three compartments of RBD. Reliability calculation for each of these blocks is as follows:

$$R(\text{A}) = r \tag{4.13}$$

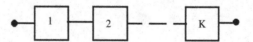

Fig. 4.8 Broadcast reliability RBD of $N \times N$ SEN

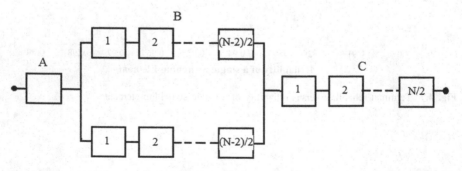

Fig. 4.9 Broadcast reliability RBD of $N \times N$ SEN+

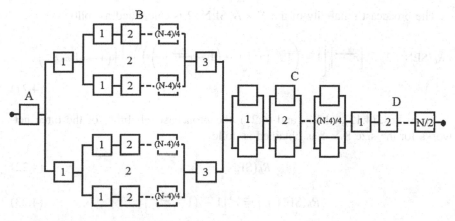

Fig. 4.10 Broadcast reliability RBD of $N \times N$ SEN+2

$$R(B) = 1 - \left(1 - r^{\frac{(N-2)}{2}}\right)^2 \qquad (4.14)$$

$$R(C) = r^{\frac{N}{2}} \qquad (4.15)$$

The broadcast reliability of the $N \times N$ SEN+ is calculated by Eq. (4.16).

$$R_b(\text{SEN}+) = r^{\frac{(N+2)}{2}}\left[1 - \left(1 - r^{\frac{(N-2)}{2}}\right)^2\right] \qquad (4.16)$$

According to Fig. 4.3, broadcast reliability RBD of $N \times N$ SEN+2 is represented in Fig. 4.10.

Actually, RBD in Fig. 4.10 is composed of four blocks, such as A, B, C, and D. Reliability for each of these blocks is given as follows:

$$R(A) = r \qquad (4.17)$$

$$R(B) = 1 - \left(1 - \left(r^2\left(1 - \left(1 - r^{\frac{(N-4)}{4}}\right)^2\right)\right)\right)^2 \qquad (4.18)$$

$$R(C) = \left(1 - (1 - r)^2\right)^{\frac{N-4}{4}} \qquad (4.19)$$

$$R(D) = r^{\frac{N}{2}} \qquad (4.20)$$

The broadcast reliability of the $N \times N$ SEN+2 is calculated as follows:

$$R_b(\text{SEN}+2) = \left[r^{\frac{(N+2)}{2}}\right]\left[1 - \left(1 - \left(r^2\left(1 - \left(1 - r^{\frac{(N-4)}{4}}\right)^2\right)\right)\right)^2\right]\left[\left(1 - (1-r)^2\right)^{\frac{N-4}{4}}\right]$$

(4.21)

From Eqs. (4.12), (4.16), and (4.21), the broadcast reliability of the three networks for the size 8×8 is obtained as follows:

$$R_b(\text{SEN}) = r^7$$

(4.22)

$$R_b(\text{SEN}+) = r^5\left[1 - (1-r^3)^2\right]$$

(4.23)

$$R_b(\text{SEN}+2) = r^5\left[1 - \left(1 - \left(r^2\left(1 - (1-r)^2\right)\right)\right)^2\right]\left[1 - (1-r)^2\right]$$

(4.24)

Table 4.2 represents the broadcast reliability results for the three networks of size 8×8. Obviously, weakest results belong to SEN while SEN+2 has the best. Figure 4.11 represents the broadcast reliability results for a convenient comparison of networks.

As it is shown in Fig. 4.11, particularly when the switch reliability is low (0.9–0.95), the broadcast reliability in SEN+2 is higher than that of other two. However, similar results are obtained in both SEN+2 and SEN+ networks; but always they have higher level of broadcast reliability compared to single-path SEN.

In general, the conclusion from these points is that although broadcast reliability in SEN+2 is higher than the other two networks, the efficiency of one extra stage is more than adding two stages in SEN; because there is dramatic increase of the reliability by adding one stage to SEN, it will be low by adding two stages.

Switching reliability	$R_b(\text{SEN})$	$R_b(\text{SEN}+)$	$R_b(\text{SEN}+2)$
0.99	0.932065	0.950151	0.950515
0.98	0.868125	0.900795	0.902115
0.97	0.807983	0.852185	0.854878
0.96	0.751447	0.804540	0.808875
0.95	0.698337	0.758041	0.764166
0.94	0.648477	0.712840	0.720805
0.93	0.601701	0.669060	0.678838
0.92	0.557846	0.62680	0.638304
0.91	0.516761	0.586136	0.599231
0.90	0.478297	0.547124	0.561638

Table 4.2 Comparison of the broadcast reliability of 8×8 SEN, SEN+, and SEN+2

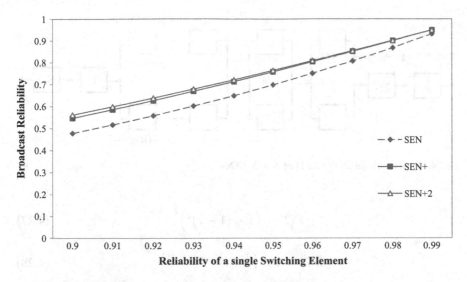

Fig. 4.11 Broadcast reliability versus reliability of a single switching element

4.2.4 Network Reliability of SEN, SEN+, and SEN+2

Network reliability defined as the probability of successful communication between all sources and all network destinations, which represents the reliability of the connections between all sources and all destinations.

So, as discussed earlier about the network structure of SEN, in this type of reliability all switches in the first and last stages are essential. Based on Fig. 4.2, network reliability RBD of $N \times N$ SEN is represented in Fig. 4.12.

Network reliability of $N \times N$ SEN is calculated by Eq. (4.25).

$$R_n(\text{SEN}) = r^{\frac{N}{2}\log_2 N} \tag{4.25}$$

Regarding to Fig. 4.2, network reliability RBD of $N \times N$ SEN+ is represented in Fig. 4.13.

The RBD in Fig. 4.13 is composed of four main blocks, such as 1, 2, 3, and 4. Reliability for each of these blocks can be obtained as follows:

$$R(1) = R(4) = r^{\frac{N}{2}} \tag{4.26}$$

Fig. 4.12 Network reliability RBD of $N \times N$ SEN

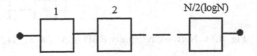

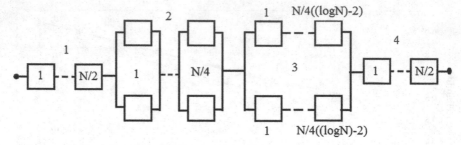

Fig. 4.13 Network reliability RBD of $N \times N$ SEN+

$$R(2) = \left(1 - (1 - r)^2\right)^{\frac{N}{4}} \tag{4.27}$$

$$R(3) = 1 - \left(1 - r^{\frac{N}{4}((\log_2 N)-2)}\right)^2 \tag{4.28}$$

The network reliability of the $N \times N$ SEN+ is obtained by Eq. (4.29).

$$R_n(\text{SEN}+) = r^N \left[\left(1 - (1 - r)^2\right)^{\frac{N}{4}}\right]\left[1 - \left(1 - r^{\frac{N}{4}((\log_2 N)-2)}\right)^2\right] \tag{4.29}$$

Based on Fig. 4.3, network reliability RBD of $N \times N$ SEN+2 is represented in Fig. 4.14. In fact, the RBD in Fig. 4.14 is composed of five main blocks, such as 1, 2, 3, 4, and 5. Reliability for each of these blocks can be obtained as follows:

$$R(1) = R(5) = r^{\frac{N}{2}} \tag{4.30}$$

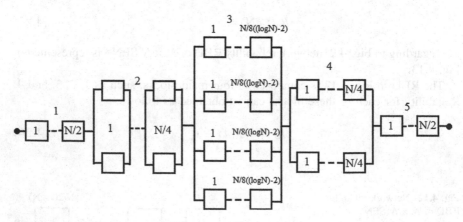

Fig. 4.14 Network reliability RBD of $N \times N$ SEN+2

$$R(2) = \left(1 - (1 - r)^2\right)^{\frac{N}{4}} \tag{4.31}$$

$$R(3) = 1 - \left(1 - r^{\frac{N}{8}((\log_2 N) - 2)}\right)^4 \tag{4.32}$$

$$R(4) = 1 - \left(1 - r^{\frac{N}{4}}\right)^2 \tag{4.33}$$

The network reliability of the $N \times N$ SEN+2 is calculated by Eq. (4.34).

$$R_n(\text{SEN} + 2) = r^N \left[\left(1 - (1 - r)^2\right)^{\frac{N}{4}}\right]\left[1 - \left(1 - r^{\frac{N}{8}((\log_2 N) - 2)}\right)^4\right]\left[1 - \left(1 - r^{\frac{N}{4}}\right)^2\right]$$

$$\tag{4.34}$$

From Eqs. (4.25), (4.29), and (4.34), the network reliability of the three networks for the size 8×8 is obtained as follows:

$$R_n(\text{SEN}) = r^{12} \tag{4.35}$$

$$R_n(\text{SEN}+) = r^8 \left[\left(1 - (1 - r)^2\right)^2\right]\left[1 - (1 - r^2)^2\right] \tag{4.36}$$

$$R_n(\text{SEN} + 2) = r^8 \left[\left(1 - (1 - r)^2\right)^2\right]\left[1 - (1 - r)^4\right]\left[1 - (1 - r^2)^2\right] \tag{4.37}$$

Network reliability results are shown in Table 4.3. In addition, the three networks, namely SEN, SEN+, and SEN+2 of size 8×8, are compared in terms of network reliability, and it is shown in Fig. 4.15.

Figure 4.15 demonstrates that as compared to SEN, always a higher level of network reliability can be achieved by means of SEN+ and SEN+2. Furthermore, SEN+ and SEN+2 give results with a very close performance with regard to

Switching reliability	$R_n(\text{SEN})$	$R_n(\text{SEN}+)$	$R_n(\text{SEN}+2)$
0.99	0.886384	0.922194	0.922194
0.98	0.784716	0.848749	0.848749
0.97	0.693842	0.779600	0.779599
0.96	0.612709	0.714663	0.714661
0.95	0.540360	0.653832	0.653828
0.94	0.475920	0.596988	0.596980
0.93	0.418596	0.543997	0.543984
0.92	0.367666	0.494716	0.494696
0.91	0.322475	0.448993	0.448963
0.90	0.282429	0.406670	0.406629

Table 4.3 Comparison of the network reliability of 8×8 SEN, SEN+, and SEN+2

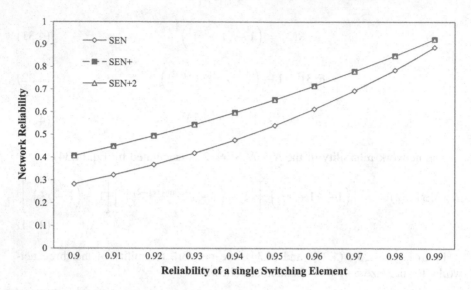

Fig. 4.15 Network reliability versus reliability of a single switching element

network reliability. According to Table 4.3, there is no doubt that that SEN+ can be a little better than SEN+2. Totally, we conclude that adding two switching stages to improve the network reliability of the SEN has much less efficiency than adding one switching stage.

While [7] reached the conclusion that SEN+ and SEN have much better reliability than the SEN+2. Here, results obtained for terminal, broadcast, and network reliability analysis show that similar reliability is achieved by both SEN+ and SEN+2 and they work much better than SEN. In compare to previous studies, it is believed that results obtained in this study are closer to reality with respect to the accuracy of reliability analysis method. Based on the results, it is also concluded the fact that adding one extra switching stage to SEN has more efficiency than adding two switching stages in terms of reliability. It is obvious that to obtain significant reliability improvement, is provided by adding one extra stage to SEN is suggested, while it is not like this about adding two stages.

Altogether, based on the obtained results, we can come to conclusion that the improvement of reliability in MINs can be achieved via adding the number of switching stages. However, this enhancement is limited and may not have efficiency in large-scale systems. So, looking for more advanced solutions is needed to improve the reliability and fault tolerance of MINs.

4.3 Designing a Multistage Interconnection Network with Fault-Tolerant Capability

As mentioned above, the two main classes of MINs are as follows: (a) single-path MINs and (b) multiple path MINs [3, 8, 10, 11]. The low hardware cost of single-path MINs like the Baseline [12], Delta [13], and Generalized Cube [14] is a good advantage for them to use in large multiprocessor systems. However, the only one path from a given source to a given destination is considered as a problem with this topology. In addition, several different nodes of source and destination use this path. Therefore, only one source–destination pair can use a particular path at the same time. Consequently, this path cannot be available for other requests. Moreover, any defect on that path leads to lose the possibility of communication. Therefore, using of fault-tolerant or multiple path networks can be considered as one of the solutions to avoid this situation or reduce it.

Pars network is a new type of regular fault-tolerant MINs. This MIN can involve in solving the blocking problem and improving some important parameters such as reliability. Therefore, Pars network is introduced in this section. In comparison with important regular MINs, namely ABN [9], ASEN [15, 16], and EGN [17], there are more number of paths from any source node to any destination node by using Pars network.

It will be demonstrated that main metrics in designing fault-tolerant MINs (i.e., cost-effectiveness, high reliability, and efficient control mechanism) can be achieved by means of Pars network.

4.3.1 Related Works

Fault tolerance as well as reliability has been two considerable issues in the field of MINs for a long time, which has resulted in the design of different topologies. As we can see in the reviewed works, two main methods to increase reliability and fault tolerance of MINs could be derived:

1. *Adding a number of stages to the network*

This methodology can provide more paths from any source node to any destination node by adding a number of stages to the network. This method is one of the primary strategies utilized to enhance reliability and fault tolerance in MINs [1, 3]. These are some of advantages of this method: (1) It can lead to an improvement in the reliability level. (2) In this approach, the extent of increasing in costs is exactly equal to which happens by adding one switching stage. In other words, this technique is known as an affordable method. (3) It is relatively simple to implement.

On the other hand, there are some defects that cause this approach to be inefficient. These disadvantages are as follows: (1) It cannot be used widely. In fact, multiple use of this approach on a given network is not efficient, as there is no guarantee for improving reliability in any number of uses of this approach on a network [7, 18]. Consequently, just we can slightly improve in reliability. (2) The length of the path from the source nodes to the destination nodes is influenced by this method, and it could be increased.

Therefore, we cannot mention the above method as a high-performance technique to improve the metrics of fault tolerance and reliability.

2. *Combining several networks in parallel on a single network*

This method can be considered as an alternative solution for the first approach [9, 12, 16, 19], and based on the results of analyses, we can get better reliability of this method compared to the first one [6]. The advantages of this approach are as follows: (1) The reliability and fault tolerance could be more increased by means of this approach compared to the first method. (2) Reducing the number of stages is possible in this method; as a result, there is reduction in the parameter of the path length from any given source to each specific destination in the network.

Despite these advantages, defects of this methodology also can be mentioned as follows: (1) Generally, by increasing the size of switching elements, the network cost increases. (2) In addition, the failure rate of a switch can be increased by increasing its size [6, 9]. (3) In this methodology, adjuvant links that are typically utilized to overcome some defects can lead to an increase in path length between each source–destination pair.

Altogether, based on the above-mentioned principles, many issues related to the second approach can be solved, by means of designing a topology that can supply the following features: (1) multiple paths between each source–destination pair must be provided by it. (2) In addition, small-sized switching elements should be used in its composition to reduce the network costs. Furthermore, practically it is shown that in switches of small sizes, the possibility of failure is less [6, 9, 18–20]. Therefore, small-size switches provide networks, which can potentially give high reliability. (3) Maintenance of performance metrics in real-time applications should be considered as much as possible, even under fault and conflict conditions.

So, introducing a topology that can meet the above features is our aim in this section. For this purpose, we used a new idea in designing high-reliability networks that is superior to previous methods in terms of fault tolerance and reliability. Additionally, real-time network performance can be achieved by this method even if there are some faulty components. In this new idea for design, the desired destination in the network is accessible by both outputs of switches on the penultimate stage. In next subsection, we will introduce its structure in detail. Moreover, the comparison of this topology with four popular fault-tolerant MINs, namely ASEN, ABN, EGN, and IEGN [10] will be done. Note that, these networks were built based on the second approach that previously described. Although ASEN, ABN, and IEGN can provide a level of reliability, they are costly because of the necessity using switches of size 3×3. On the contrary, EGN is cheap by using small-size 2×2 switches, but less number of paths from any given source to any destination can be provided by it in comparison with ASEN and ABN. Moreover, it is considerable that [10] demonstrates significant advantage of the IEGN against the networks of ASEN, ABN, and EGN. Nevertheless, in [10] similar to [21, 22], reliability is considered to be the same for all switching components, even components of various sizes. However, practically less reliability is provided by larger size switches in compare to smaller size switches [6, 9, 18–20]. Therefore, according to this reasonable assumption, networks of ASEN, ABN, and IEGN will not be able to get adequate reliability in compare to the networks such as EGN. This is due to the fact that these networks are using switches of size 3×3 while EGN uses small-size 2×2 switches.

4.3.2 Pars Network Structure

An 8×8 and 16×16 Pars network is shown in Figs. 4.16 and 4.17, respectively (in Fig. 4.17, the connections are illustrated only to the source 0000 for better clarity). A Pars network of size $N \times N$ is made of $((\log_2 N) - 1)$ stages of $\left(\frac{3N}{4}\right)$ switching elements. All switches are of small size of 2×2 that led to reduction the

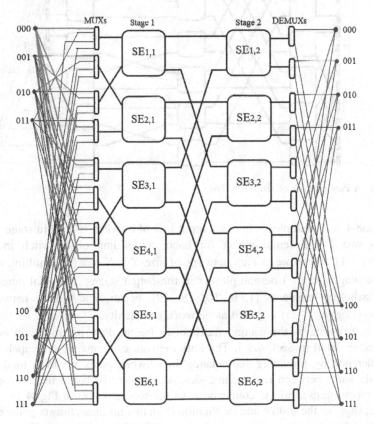

Fig. 4.16 A Pars network of size 8×8

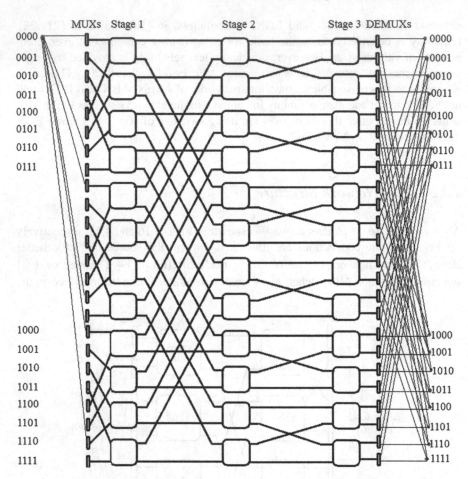

Fig. 4.17 A Pars network of size 16×16

costs. One 4×1 multiplexer for each input link of a switch is used in stage 1, and there is one 1×2 demultiplexer for each output link of a switch in stage $((\log_2 N) - 1)$. Therefore, a Pars network of size $N \times N$ has $\left(\frac{3N}{2}\right)$ multiplexers in the input stage and $\left(\frac{3N}{2}\right)$ demultiplexers in the output stages. The total number of 2×2 switching elements [1, 6, 10, 18, 19], of an $N \times N$ Pars network is $\left(\frac{3N}{4}\right)\left(3 + ((\log_2 N) - 1)\right)$ is called as network complexity.

The self-routing as the routing mechanism for the Pars network will be also discussed in detail in Sect. 4.3.3. The Pars network can enhance the capability of fault tolerance by providing redundancy in network paths up to a total number of six paths between each source–destination pair. To better understand, different colors are used to the connections to the source = 000 in Fig. 4.16. As an instance, suppose the source and destination 0. In this instance, the six paths can be listed as follows: $000 \rightarrow SE_{1,1} \rightarrow SE_{1,2} \rightarrow 000$, $000 \rightarrow SE_{1,1} \rightarrow SE_{3,2} \rightarrow 000$,

$000 \rightarrow SE_{3,1} \rightarrow SE_{1,2} \rightarrow 000$, $000 \rightarrow SE_{3,1} \rightarrow SE_{5,2} \rightarrow 000$, $000 \rightarrow SE_{5,1} \rightarrow$ $SE_{3,2} \rightarrow 000$, $000 \rightarrow SE_{5,1} \rightarrow SE_{5,2} \rightarrow 000$.

Here the question is by using more than two stages in Pars of larger sizes what connection template is considered for interconnecting stages? Typically, each source is connected to six multiplexers and each destination is connected to three demultiplexers in the Pars network. Suppose we divide the number of multiplexers into three groups so that each group includes $\left(\frac{3N}{6}\right)$ multiplexers. Therefore, any given source is connected to two multiplexers in each group. In addition, each of these two multiplexes ends in a certain group of destinations. If Destination $= d_1 d_2 \ldots d_{\log_2 N}$, one of these two multiplexers is considered for communication according to bit d_2. Another point is that if demultiplexers are divided into three groups so that each group contains $\left(\frac{3N}{6}\right)$ demultiplexers, then any given destination will be connected to one demultiplexer in each of these groups. Take to consideration that these demultiplexers are selected to connect to destinations according to bit d_2. Moreover, it should be explained that if the switches located at each stage are divided into two groups, then each group results in a particular group of destinations according to bit d_2. For a better understanding of these discussions, for instance, we consider the source 000 in Fig. 4.16. As discussed, six multiplexers should be connected to the source 000. The six multiplexers should be chosen to connect in respect to these points: (a) by considering multiplexers in three groups, each containing $\left(\frac{3N}{6}\right)$ multiplexers (i.e., four multiplexers in Fig. 4.16 and eight multiplexers in Fig. 4.17), then any given source is connected to two multiplexers in each of these groups. As it is shown, both Figs. 4.16 and 4.17 have mentioned this condition. (b) If the switches located in each stage are categorized into two groups, then each of these groups will end in a specific group of destinations. Based on this fact, in each of the multiplexers, the source should be connected to two multiplexers in such a way that each multiplexer has access to a specific group of destinations. To better understanding this discussion, consider Fig. 4.16. Six switches are available in stage 1. Among these switches, $SE_{1,1}$, $SE_{3,1}$, and $SE_{5,1}$ have access to destinations 000, 001, 100, and 101 $(d_2 = 0)$. Also, switches $SE_{2,1}$, $SE_{4,1}$, and $SE_{6,1}$ have access to destinations 010, 011, 110, and 111 $(d_2 = 1)$. Therefore, in each of the three groups of multiplexers, the source 000 must be linked to two multiplexers in such a way that selected two multiplexers are connected to two switches that can be terminated in various groups of destinations. As it is shown, both Figs. 4.16 and 4.17 have the same condition in this case.

In general, it should be noted that if the switches in each stage are divided into three groups so that each group contains $\left(\frac{3N}{12}\right)$ switches. The first $\left(\frac{3N}{24}\right)$ switches of this number of switches will have access to the destinations that bit d_2 in them is zero. In addition, the second $\left(\frac{3N}{24}\right)$ switches will have access to the destinations that bit d_2 in them is one.

Considering the above discussion will be useful for understanding the design of the Pars network for various network sizes. However, we also suggest a suitable method for designing larger network sizes that can solve the issue of the communication template utilized between stages in larger size networks with more than two stages. It should be considered that communication template used between stages in $(N/2) \times (N/2)$ Pars network can be used to derive communication template that should be utilized between stages in $N \times N$ Pars network. The communication template utilized between stages 1 to $(\log_2 N) - 1$ of $(N/2) \times (N/2)$ Pars network can be considered to design the communication template that should be utilized between stages 1 to $(\log_2 N) - 2$ in an $N \times N$ Pars network. For this purpose, these points should be considered: In each stage of the networks, the switching elements are divided into groups containing $\left(\frac{3N}{24}\right)$ switches. Each of these switch groups in the $N \times N$ Pars network, including $\left(\frac{3N}{24}\right)$ switches per group, must comply with the communication template used by the $\left(\frac{3N}{24}\right)$ corresponding switches in the $(N/2) \times (N/2)$ Pars network. To better understanding, an instance is examined. Suppose the 8×8 Pars network (Fig. 4.16), $\left(\frac{3N}{24}\right)$ is equal to one in this network. As a result, the switches at each stage are considered in single-member groups. Now, suppose the 16×16 Pars network (Fig. 4.17), $\left(\frac{3N}{24}\right)$ is equal to two in this network. As a result, the switches at each stage are considered in two-member groups. The communication template utilized via the two-member groups of 16×16 Pars network should follow the communication pattern utilized by the single-member groups of 8×8 Pars network. As an example, between stages 1 and 2 in the 16×16 Pars network, the switches of $SE_{1,1}$ and $SE_{2,1}$ (the first digit of the legend represents the number of the switch and the second one indicates the number of the stage) should follow the communication template utilized by the $SE_{1,1}$ of 8×8 Pars network. As a result, since the $SE_{1,1}$ from 8×8 Pars is linked to $SE_{1,2}$ and $SE_{3,2}$ of stage 2, the $SE_{1,1}$ and $SE_{2,1}$ of 16×16 Pars should be linked to the corresponding switches of $SE_{1,2}$, $SE_{2,2}$ (these switches are corresponding to $SE_{1,2}$ of 8×8 Pars), $SE_{5,2}$, $SE_{6,2}$ (these switches are corresponding to $SE_{3,2}$ of 8×8 Pars) of 16×16 Pars, in such a way that the first switch $SE_{1,1}$ of 16×16 Pars should be linked to the first switches $SE_{1,2}$ and $SE_{5,2}$ as well as the second switch $SE_{2,1}$ of 16×16 Pars should be linked to the second switches $SE_{2,2}$ and $SE_{6,2}$. To better understanding this discussion, the switching elements in 8×8 Pars network as well as their corresponding switches in the 16×16 Pars network are summarized in Table 4.4.

Above method can be used to determine the pattern of the communications between stages one through $(\log_2 N) - 2$ of $N \times N$ Pars network based on the pattern of the communications between stages one to $(\log_2 N) - 1$ of $(N/2) \times (N/2)$ Pars network. Nevertheless, the communication template that must be used between stages $(\log_2 N) - 2$ and $(\log_2 N) - 1$ of $N \times N$ Pars has not been determined yet. To determine this item, the previously mentioned issue guides that the switches located in any stage divided into three groups and each group contains $\left(\frac{3N}{12}\right)$ switches. Destinations that with bit d_2 equal to 0 could be resulted by the first $\left(\frac{3N}{24}\right)$ switches of these $\left(\frac{3N}{12}\right)$ switches. Also, by the next $\left(\frac{3N}{24}\right)$ switches of these $\left(\frac{3N}{12}\right)$ switches, we can get destinations bit d_2 equal to 1. So, the organization of the connection pattern used among stages $(\log_2 N) - 2$ and $(\log_2 N) - 1$ of $N \times N$ Pars should be in such a way as to adhere to this issue.

Table 4.4 Corresponding switches in the 8×8 and 16×16 Pars network

Switches in stage 1 of 8×8 Pars	Corresponding switches in stage 1 of 16×16 Pars	Switches in stage 2 of 8×8 Pars	Corresponding switches in stage 2 of 16×16 Pars
$SE_{1,1}$	$(SE_{1,1}, SE_{2,1})$	$SE_{1,2}$	$(SE_{1,2}, SE_{2,2})$
$SE_{2,1}$	$(SE_{3,1}, SE_{4,1})$	$SE_{2,2}$	$(SE_{3,2}, SE_{4,2})$
$SE_{3,1}$	$(SE_{5,1}, SE_{6,1})$	$SE_{3,2}$	$(SE_{5,2}, SE_{6,2})$
$SE_{4,1}$	$(SE_{7,1}, SE_{8,1})$	$SE_{4,2}$	$(SE_{7,2}, SE_{8,2})$
$SE_{5,1}$	$(SE_{9,1}, SE_{10,1})$	$SE_{5,2}$	$(SE_{9,2}, SE_{10,2})$
$SE_{6,1}$	$(SE_{11,1}, SE_{12,1})$	$SE_{6,2}$	$(SE_{11,2}, SE_{12,2})$

The final point is that the routing algorithm (discussed in Sect. 4.3.3) to prove the perfect design of can examine the Pars network. Here, the main point is that the routing mechanism should be able to be implemented for any optional size of the Pars network. As a result, the evaluation of the routing algorithm on the network can be used as a benchmark for determining the design accuracy of the connection pattern used among stages.

The considerable point is that there is a special feature in Pars network structure that we cannot see in previous studies. This feature is that we can get the desired destination by both outputs of the switch in the penultimate stages. We called it as *perfect connection* feature. For example, suppose the source 000 and the destination 000. Figure 4.18 shows the perfect connection feature for this example. As it is shown, we got destination 000 by both outputs of $SE_{1,1}$, $SE_{3,1}$, and $SE_{5,1}$. Actually, considering this feature, two successor switches are considered in stage 2 per each switch in stage 1. If the first successor switch is not available due to some faults or blocking the request, this structure can provide the possibility of using the second successor switch very quickly.

Some of the advantages of the Pars network can be listed as follows: (1) Its cost has dropped due to use of switching elements of size 2×2 (which is the minimum size) in its composition. Furthermore, there is less likely of failure of small-size switches [6, 9, 18–20] as it has been proven practically. Therefore, high reliability can be achieved by means of networks that are made of small-size switches. (2) It is powerful to correct the fault, since it has guaranteed six different paths between each pair of source–destination. (3) Even in the presence of fault, it has advantages like reducing blockage, reducing delays, and real-time performance because it includes an exact feature called as *perfect connection*.

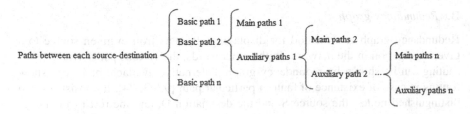

Fig. 4.18 Different types of paths in MINs

The confirmation of the above claims will be provided by Pars network performance analysis, which will be discussed in Sect. 4.3.4.

4.3.3 Proposed Routing Scheme for Pars Network

We will study these topics in this subsection: In section A, a classification of network paths will be discussed. In section B, redundancy graph for Pars network will be presented. In section C, the structure of the switching elements is investigated. At last, in section D, the self-routing for Pars network will be studied.

A. Different types of paths in MINs

A classification of different paths in MINs, which will be used in later discussions, is presented in this subsection.

For more exact investigation, the paths from any given source to each given destination are classified into major groups called base paths, while any of them in turn is classified as main and auxiliary paths. By multiplexers between any source–destination, basic paths (BPs) are selected. The basic path 1 at first is tried by every request, and then if there is failure in this procedure, other basic paths will be tried in the same way. In ordinary conditions, main paths (MPs) are used and have shorter length compare to auxiliary paths. However, in case of being busy or presence of a fault that can cause main paths to be out of availability, auxiliary paths (APs) which are longer in length than main ones are used. Primitively shorter paths are used in an auxiliary path that has a set of paths with different lengths. Therefore, as presented in Fig. 4.18, auxiliary paths can be categorized in main and auxiliary paths called MP_n and AP_n that $n \in [2, +\infty)$.

To better understand this classification, Fig. 4.19 shows an ASEN [15, 16] of size 8×8.

For the source = 000 and destination = 101, different types of paths are listed below:

Basic path 1:
Main path 1: $000 \rightarrow SE_{0,1} \rightarrow SE_{2,2} \rightarrow 101$
Auxiliary path 1: $000 \rightarrow SE_{0,1} \rightarrow SE_{1,1} \rightarrow SE_{3,2} \rightarrow 101$
Basic path 2:
Main path 1: $000 \rightarrow SE_{2,1} \rightarrow SE_{2,2} \rightarrow 101$
Auxiliary path 1: $000 \rightarrow SE_{2,1} \rightarrow SE_{3,1} \rightarrow SE_{3,2} \rightarrow 101$

B. Redundancy graph

Redundancy graph is a method for displaying all paths from a given source to a given destination in the network. In the other words, the various paths available for routing can be shown by redundancy graph. Alternative available paths can show routing in case of existence of fault in particular path [9, 23, 24]. It consists of two distinguished nodes, the source S and the destination D, and the rest of the nodes

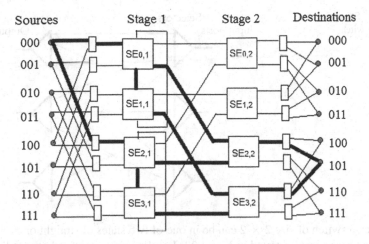

Fig. 4.19 Classification of existing paths in MINs

correspond to the switches along the paths between S and D. Figure 4.20 shows the redundancy graphs of Pars network.

As Fig. 4.20 shows, three basic paths have been provided from any given source to any given destination in which two main paths compose each of them. In fact, all of these paths have the same length. Therefore, Pars network can act with appropriate speed in response to requests, even in faulty conditions.

C. *Designing switching elements*

MINs are organized in a number of stages that each of these stages are made up of several small-size crossbar switches. A crossbar switch suitable for distributed control in the environment a MIN is shown in Fig. 4.21 [8]. Each incoming packet to the switch contains a tag that indicates the destination of the packet. The selector of the input port checks this tag, and then it considers the appropriate output port based on it. If the arbiter of this output port is free, then the connection is established. However, if it is busy, then the connection could not be established. It should be noted that all selectors can work concurrently and asynchronously.

Fig. 4.20 Redundancy graph
for Pars network

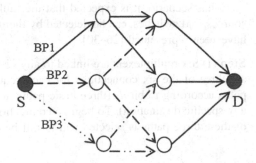

Fig. 4.21 Structure of a
crossbar switch

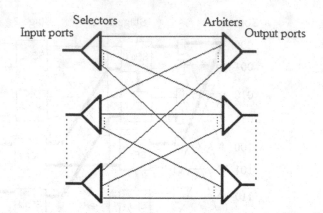

In fact, a switch of size 2×2 can be in one of two states of straight or exchange. These two cases are illustrated in Fig. 4.22. Let the upper input and output lines be labeled by i and the lower input and output lines be labeled by j, (1) straight-input i to output i, input j to output j and (2) exchange-input i to output j, input j to output i.

D. *Routing mechanism in Pars network*

Routing on the Pars network can be implemented in the form of self-routing. In self-routing mechanism, the switching elements set themselves by examination of the destinations of their input data. There is no need to compute for the routing decision in a self-routing procedure. In addition, it does not need to hold routing tables inside switch fabric. Also, in a self-routing algorithm, very simple hardware can do routing decisions at a very short time [8, 25].

Binary digits, which are form the Routing tag, control the connection through different stages of the path from input to the output. Let the source S and destination D is represented in binary as:

$$S = s_1 s_2 \ldots s_n$$

$$D = d_1 d_2 \ldots d_n$$

and $n = \log_2 N$.

Self-routing algorithm in Pars network will be as follows:

In this scenario, it is expected that the faults of the switches, which connect to sources, and switches, can be detected by them. So far, several detecting techniques have been represented [26–30].

Step (1) Six multiplexers are linked to any given source in the Pars network. There are three of the six connections for each request that are considered as the basic paths according to bit d_2 (three basic paths are provided from any given source to any specific destination). To begin a connection from a source to a destination, one of these three paths is selected and it will be considered as the basic path 1.

Fig. 4.22 Structure of a crossbar switch

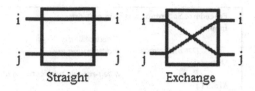

Straight   Exchange

Step (2) (i) *If there are no faults in the network:* Bits of d_2 to d_n are considered for routing in the middle stages between multiplexers and demultiplexers. For each request appeared on each switch, bit 0 means to establish a connection with the upper output. Also, bit 1 means to establish a connection with the lower output. As an instance, this issue is illustrated in Fig. 4.23.

(ii) If there are *faults* in the network: Initially, basic path 1 is considered to perform any request. However, it is possible that required first switch in the first stage to start the basic path 1 is faulty or inaccessible. In this case, the next path is used.

On basic path 1, if the required switch on the next stage is faulty, then the request is sent to the other output of the switch in the current stage. Therefore, the new successor switch can use the same routing tag to continue. Now, assume that this new switch on the next stage is also unavailable. In this case, the next basic path should be used with the same previous procedure. Finally, if all possible paths are unavailable on all basic paths, then the request will collapse.

Step (3) Finally, in multiplexers, the request is sent to the upper or lower destination based on the most significant bit of the tag (i.e., d_1).

According to the mentioned explanation, the self-routing algorithm is illustrated in Fig. 4.24.

A permutation is defined as a two-row matrix bounded by parentheses. The top row is the list of sources, and the bottom row is the list of destinations [10, 31]. Figure 4.25 shows a successful Pars network routing procedure for permutation

$$P = \begin{pmatrix} 0 & 1 & 2 & 3 & 4 & 5 & 6 & 7 \\ 1 & 6 & 4 & 7 & 2 & 3 & 5 & 0 \end{pmatrix}.$$

4.3.4 Performance Analysis

MINs analysis can be performed from different aspects. Terminal reliability, fault-tolerant, cost, and permutation capability are those aspects we will analyze the

Fig. 4.23 Self-routing in each switch

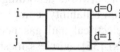

If d=0 then input is connected to output i
If d=1 then input is connected to output j

```
Procedure: Self-routing of Pars network
Inputs:
   -  N                         // network size
   -  D = d₁d₂...d_logN         // destination tag
Outputs:
   -  P                         // self-routing path between a source – destination pair
Main ( )
{
   If (condition is non-faulty) then
      -  three available basic paths are identified by input bit d₂.
      -  one of the basic paths is selected to initiate communication.
      -  bits d₂ to d_logN are used for self-routing between intermediate stages 1 to [(logN)-1]
      -  from the demultiplexer, the request is routed to the upper or lower destinations based on the input bit d₁.
   Else
      P = FIND_PATH ( );
      Return P;
}
----------------------------------------
Procedure FIND_PATH ( )
Outputs:
   - P                          // self-routing path between a source – destination pair
{
   For each basic_path ( i )
   If ( is non-faulty)
      For each output c of path (i).stage (1)
         If available (c)
         {
            P = c...d_logN d₁;
            Return (P);
         }
   request.drop ( );
}
```

Fig. 4.24 Self-routing algorithm for Pars network

networks in this section. Reasons for choosing these parameters are as follows: Important role of reliability in determining system performance is a major concern in MINs. Also, continuous and long-period operation of the network is possible by fault-tolerant capability. The reasonable cost is the next considerable metric. Finally, arrival rate, mean time to arrival, response time, and efficiency are the useful information about MINs that is provided by permutation capability, and each of them will be studied in the section D.

A. *Terminal reliability*

Generally, reliability can provide a system to perform and maintain its activities in conventional conditions, as well as adverse or accidental situations. Therefore, it is the most immediate parameter for each efficient network in the domain of inter-connection networks according to many researches [1, 6, 7, 9–11, 17–20, 24].

Based on reliability, complex network systems are as follows: lifeline networks such as electrical and gas networks [32–34], wireless mobile ad hoc networks (MANETs) [35], wireless mesh networks [36–39], wireless sensor networks [40, 41], nano-sensor based on nano-wired networks [42], social networks [43],

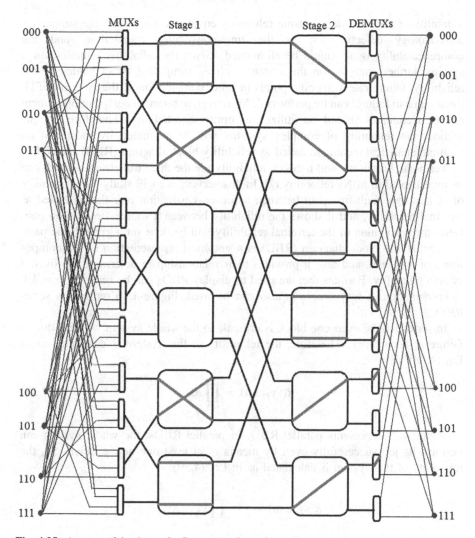

Fig. 4.25 A successful scheme for Pars network routing

stochastic-flow manufacturing networks (SMNs) [44], and interconnection networks [1, 6, 10, 18].

Researches demonstrated that simulation or analytical models can be used in reliability investigation of the complex networks. Despite easy implementation of simulation-based approaches, their effectiveness has some restrictions. For example, to perform a comprehensive study, the large number of performed simulations should be provided that it would be so time-consuming. Additionally, the analytical methods can provide a wide range of results compared to simulation. Clearly, simulation is more simple and so analytical methods have been avoided due to their complexity. Analytical methods can accomplish an exact solution for computing the reliability of a system due to using reliability equations. Therefore, the simulation methodology defects such as the time-consuming calculations and the non-repeatability issue should be eliminated. Given the reliability equation for a system, further analyses on the system such as computing exact values of the reliability, failure rate at specific points in time, computation of the system MTTF (mean time to failure) can be performed. Moreover, to promote design improvement efforts, techniques should be utilized to optimize reliability. Therefore, in this section, the reliability of complex systems will be evaluated by means of an accurate analytical technique called as reliability block diagram (RBD) method.

Terminal, broadcast, and network reliability are the three dimensions which can be investigated in MINs reliability [7]. In this section, we will study the probability of at least one fault-free path between a source–destination pair that is called as terminal reliability, and it shows the reliability between a source–destination pair. Here, the calculation of the terminal reliability will be done on a given basic path.

A reliability block diagram (RBD) is a graphical representation of the components of a system, and also it provides their relationship to determine the overall system reliability. Formats that are used to display RBDs can be series or parallel. A combination of two simple is also can be used. Figure 4.26 represents series RBD.

In series RBDs, even one block is critical, so the whole system will fail due to failure of each block. Therefore, the reliability of this system is calculated as in Eq. (4.38).

$$R(\text{system}) = \prod_{i=1}^{n} Ri \tag{4.38}$$

Figure 4.27 represents parallel RBD. In parallel RBDs, the whole system can perform its job successfully even by means of at least one active block. So, the reliability of this system is calculated as in Eq. (4.39).

$$R(\text{system}) = 1 - \prod_{i=1}^{n} (1 - Ri) \tag{4.39}$$

In terminal reliability analysis, it will be assumed that the r is the probability of a 2×1 multiplexer (or a 1×2 demultiplexer) being operational. In this case, given the number of gates per each switching element, the probability of it being operational can be computed based on r. It is supposed that the more complex hardware of a component, the more number of gates and vice versa [9, 19]. According to above discussions, the reliability of each switching components is calculated as follows:

Fig. 4.26 Series RBD

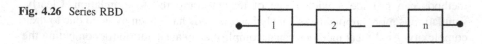

Fig. 4.27 Parallel RBD

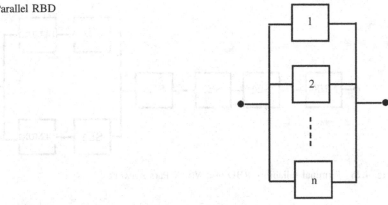

$$r_{\text{MUX2}} = r_{\text{DEMUX2}} = r \tag{4.40}$$

$$r_{\text{MUX4}} = r_{\text{SE2}} = r^2 \tag{4.41}$$

$$r_{\text{SE3}} = r^{\frac{9}{2}} \tag{4.42}$$

Figure 4.28 shows terminal reliability RBD of Pars network of size $N \times N$ is depicted. Note that the characteristics of each network must be represented by RBD. A switching element of size $n \times n$ is shown as SE_n, and a $n \times 1$ multiplexer and $1 \times n$ demultiplexer are shown as MUX_n and DEMUX_n, respectively. The terminal reliability of the $N \times N$ Pars network is calculated as in Eq. (4.43).

$$R_t(\text{Pars}) = (r_{\text{MUX4}})(r_{\text{SE2}})^{(\log_2 N)-2}\left(1 - (1 - (r_{\text{SE2}}r_{\text{DEMUX2}}))^2\right) \tag{4.43}$$

The terminal reliability for four networks of ABN, ASEN, EGN, and IEGN of size $N \times N$ is calculated as Eqs. (4.44–4.47).

$$R_t(\text{ABN}) = (r_{\text{MUX4}})(r_{\text{SE3}})^{(\log_2 N)-3}(1 - ((1 - (r_{\text{SE2}}r_{\text{DEMUX2}}))(1 - (r_{\text{SE3}}r_{\text{SE2}}r_{\text{DEMUX2}})))) \tag{4.44}$$

$$R_t(\text{ASEN}) = (r_{\text{MUX2}})(r_{\text{SE3}})^{(\log_2 N)-2}(1 - ((1 - (r_{\text{SE2}}r_{\text{DEMUX2}}))(1 - (r_{\text{SE3}}r_{\text{SE2}}r_{\text{DEMUX2}})))) \tag{4.45}$$

$$R_t(\text{EGN}) = (r_{\text{MUX2}})(r_{\text{SE2}})^{(\log_2 N)-1}(r_{\text{DEMUX2}}) \tag{4.46}$$

$$\begin{aligned}
R_t(\text{IEGN}) = &\left((r_{\text{MUX2}})(r_{\text{SE3}})^{(\log_2 N)-2}\right)(r_{\text{SE3}}(1 - (1 - r_{\text{SE3}}r_{\text{DEMUX2}})(1 \\
&- \left(1 - (1 - r_{\text{SE3}}r_{\text{DEMUX2}})\left(1 - (r_{\text{SE3}})^2 r_{\text{DEMUX2}}\right)\right))) \\
&+ (1 - r_{\text{SE3}})\left(1 - (1 - r_{\text{SE3}}r_{\text{DEMUX2}})\left(1 - (r_{\text{SE3}})^2 r_{\text{DEMUX2}}\right)\right))
\end{aligned} \tag{4.47}$$

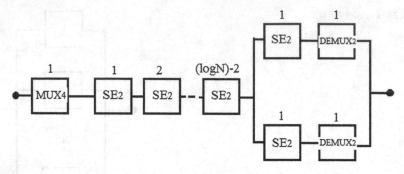

Fig. 4.28 Terminal reliability RBD of a $N \times N$ Pars network

Moreover, according to Eqs. (4.40–4.42), we have:

$$R_t(\text{Pars}) = r^{2(\log_2 N)-2}\left(1 - (1 - r^3)^2\right) \tag{4.48}$$

$$R_t(\text{ABN}) = r^{\frac{9}{2}(\log_2 N)-\frac{23}{2}}\left(1 - \left((1 - r^3)\left(1 - r^{\frac{15}{2}}\right)\right)\right) \tag{4.49}$$

$$R_t(\text{ASEN}) = r^{\frac{9}{2}(\log_2 N)-8}\left(1 - \left((1 - r^3)\left(1 - r^{\frac{15}{2}}\right)\right)\right) \tag{4.50}$$

$$R_t(\text{EGN}) = r^{2(\log_2 N)} \tag{4.51}$$

$$R_t(\text{IEGN}) = \left(r^{\frac{9}{2}(\log_2 N)-8}\right)\left(r^{\frac{9}{2}}\left(1 - \left(1 - r^{\frac{11}{2}}\right)\left(1 - \left(1 - \left(1 - r^{\frac{11}{2}}\right)(1 - r^{10})\right)\right)\right)\right.$$
$$\left. + \left(1 - r^{\frac{9}{2}}\right)\left(1 - \left(1 - r^{\frac{11}{2}}\right)(1 - r^{10})\right)\right) \tag{4.52}$$

According to (4.48–4.52), terminal reliability of these networks for the size 16×16 is calculated as follows:

$$R_t(\text{Pars}) = r^6\left(1 - (1 - r^3)^2\right) \tag{4.53}$$

$$R_t(\text{ABN}) = r^{\frac{13}{2}}\left(1 - \left((1 - r^3)\left(1 - r^{\frac{15}{2}}\right)\right)\right) \tag{4.54}$$

$$R_t(\text{ASEN}) = r^{10}\left(1 - \left((1 - r^3)\left(1 - r^{\frac{15}{2}}\right)\right)\right) \tag{4.55}$$

$$R_t(\text{EGN}) = r^8 \tag{4.56}$$

$$R_t(\text{IEGN}) = \left(r^{10}\right)\left(r^{\frac{9}{2}}\left(1 - \left(1 - r^{\frac{11}{2}}\right)\left(1 - \left(1 - \left(1 - r^{\frac{11}{2}}\right)\left(1 - r^{10}\right)\right)\right)\right)\right.$$
$$\left. + \left(1 - r^{\frac{9}{2}}\right)\left(1 - \left(1 - r^{\frac{11}{2}}\right)\left(1 - r^{10}\right)\right)\right) \tag{4.57}$$

Table 4.5 summarizes the terminal reliability results for the five networks of Pars, ABN, ASEN, EGN, and IEGN of size 16×16.

Furthermore, the reliability improvement of Pars network compared to the other four networks is represented in Fig. 4.29. In Fig. 4.29, it is demonstrated that compared to the other four networks, higher terminal reliability can be achieved by Pars network, especially when the switching components have low reliability. These results represent the fact that using Pars network provides better performance compared to other four networks, even when suitable conditions are not available (low reliability of the switch).

Although Fig. 4.29 provides beneficial information for the evaluation of the networks, investigating reliability of them in large-size networks is valuable. Therefore, Figs. 4.30, 4.31, and 4.32 represent the results of terminal reliability analysis for Pars network, ABN, ASEN, EGN, and IEGN as a function of network size according to Eqs. (4.48–4.52).

The terminal reliability as a function of network size for low switch reliability is shown in Fig. 4.30. Based on this figure, the highest terminal reliability is obtained by the Pars compared to the other networks for all network sizes. Also, achieved results confirm that even when switching components have low reliability as an undesirable scenarios, Pars is superior in terms of terminal reliability.

Figure 4.31 illustrates the results of terminal reliability analysis for different network sizes from 32 to 4096 for modest-level switch reliability ($r = 0.95$). Again, Pars is superior in terms of reliability compared to the other three networks for different network sizes according to the results. Additionally, when we compare Figs. 4.31 and 4.32, the network size and the terminal reliability of all the networks are inversely proportional. However, increase in the switch reliability from 0.9 (Fig. 4.30) to 0.95 (Fig. 4.31) leads to improve conditions and slow reduction of reliability of the

Table 4.5 Comparison of the terminal reliability for network size 16×16

r	$R_t(\text{Pars})$	$R_t(\text{ABN})$	$R_t(\text{EGN})$	$R_t(\text{ASEN})$	$R_t(\text{IEGN})$
0.99	0.940650	0.934740	0.922744	0.902431	0.903937
0.98	0.882779	0.869688	0.850763	0.810316	0.814197
0.97	0.826620	0.805751	0.783743	0.724272	0.729582
0.96	0.772358	0.743628	0.721389	0.644622	0.649813
0.95	0.720139	0.683845	0.663420	0.571466	0.57504
0.94	0.670070	0.626782	0.609569	0.504736	0.505542
0.93	0.622226	0.572697	0.559582	0.444238	0.44157
0.92	0.576656	0.521752	0.513219	0.389692	0.383272
0.91	0.533384	0.474025	0.470252	0.340758	0.33066
0.90	0.492211	0.429535	0.430467	0.297062	0.283623

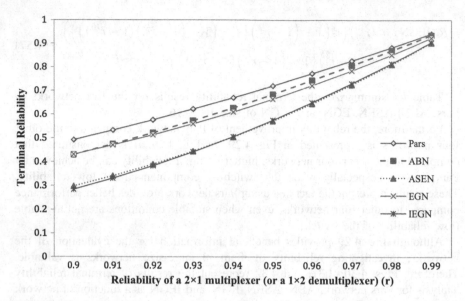

Fig. 4.29 Terminal reliability in each basic path versus reliability of a 2×1 multiplexer (or a 1×2 demultiplexer) (r)

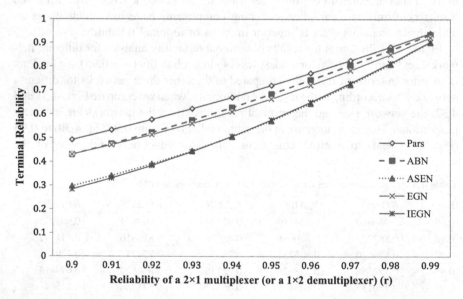

Fig. 4.30 Terminal reliability in each basic path as a function of network size for low switch reliability $(r = 0.9)$

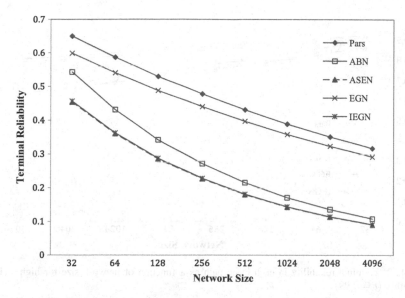

Fig. 4.31 Terminal reliability in each basic path as a function of network size for middle switch reliability ($r = 0.95$)

networks. That is, the entire network will be improved by the improvement in switching component reliability according to these two latter figures.

Moreover, the results of terminal reliability analysis as a function of network size for high switch reliability (0.99) are represented in Fig. 4.32. As the figure demonstrates, terminal reliabilities of the networks approach to each other in case of switch reliability of 0.99, compared to Figs. 4.30 and 4.31. However, Pars achieves undeniable advantages compared to other networks of ABN, ASEN, EGN, and IEGN in terms of terminal reliability. Overall, based on Figs. 4.30, 4.31, and 4.32, Pars improves other networks in terms of reliability for three switch reliability configurations (i.e., low, middle, and high) and different network sizes.

In most systems, mean time to failure (MTTF) is another important parameter that is used to measure the reliability. Within interconnection networks, one of the performance metrics is this parameter [1, 6, 9, 17, 19]. As a result, the key parameter is used to the networks to achieve a comprehensive assessment. In this section, like [9, 11, 17, 19, 20], it is assumed that the times to failure of the switching elements is based on an exponential distribution. In addition, an acceptable value for λ is about 10^{-6} per hour [9, 19, 24]. The MTTF is calculated using Eq. (4.58).

$$\text{MTTF} = \int_0^\infty R(t)\mathrm{d}t \tag{4.58}$$

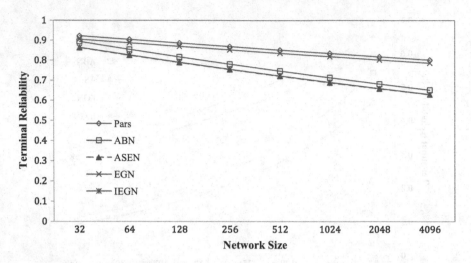

Fig. 4.32 Terminal reliability in each basic path as a function of network size for high switch reliability ($r = 0.99$)

According to Eq. (4.58), we have:

$$\text{MTTF(Pars)} = \int_0^\infty \left(e^{-(2(\log_2 N)-2)\lambda t} \left(1 - \left(1 - e^{-3\lambda t} \right)^2 \right) \right) dt \qquad (4.59)$$

$$\text{MTTF(ABN)} = \int_0^\infty \left(e^{-\left(\frac{9}{2}(\log_2 N)-\frac{23}{2}\right)\lambda t} \left(1 - \left(\left(1 - e^{-3\lambda t} \right) \left(1 - e^{-\frac{15}{2}\lambda t} \right) \right) \right) \right) dt$$
$$(4.60)$$

$$\text{MTTF(ASEN)} = \int_0^\infty \left(e^{-\left(\frac{9}{2}(\log_2 N)-8\right)\lambda t} \left(1 - \left(\left(1 - e^{-3\lambda t} \right) \left(1 - e^{-\frac{15}{2}\lambda t} \right) \right) \right) \right) dt$$
$$(4.61)$$

$$\text{MTTF(EGN)} = \int_0^\infty \left(e^{-(2(\log_2 N))\lambda t} \right) dt \qquad (4.62)$$

$$\text{MTTF(IEGN)} = \int_0^\infty \left(\left(e^{-\left(\frac{9}{2}(\log_2 N)-8\right)\lambda t} \right) \right.$$
$$\left(e^{-\frac{9}{2}\lambda t} \left(1 - \left(1 - e^{-\frac{11}{2}\lambda t} \right) \left(1 - \left(1 - e^{-\frac{11}{2}\lambda t} \right) \left(1 - e^{-10\lambda t} \right) \right) \right) \right)$$
$$\left. + \left(1 - e^{-\frac{9}{2}\lambda t} \right) \left(1 - \left(1 - e^{-\frac{11}{2}\lambda t} \right) \left(1 - e^{-10\lambda t} \right) \right) \right) dt \qquad (4.63)$$

Figure 4.33 represents the results of MTTF analysis as a function of network size according to the above equations. Considering the figure, Pars and ASEN got the best and the worst results, respectively. Based on this figure, a significant advantage in terms of MTTF compared to the other four networks is seen in Pars network.

Mostly, obtaining a failure distribution of the entire system based on the failure distribution of its components is the initial aim in performance and reliability analysis of the fault-tolerant systems. System failure rate can be used as the parameter to study these cases. In fact, failure rate of a system is used to indicate the proneness to failure of the system after time t has elapsed. Therefore, in this section we will analyze this key parameter to conduct a deeper performance analysis of the networks. The system failure rate, denoted λ_s, can be calculated by the following equation [4, 45, 46].

$$\lambda_s = -\frac{1}{R(t)}\frac{d(R(t))}{dt} \tag{4.64}$$

According to Eq. (4.64), we have:

$$\lambda_s(\text{Pars}) = \frac{e^{t\lambda(2\log_2 N-2)}\left(\dfrac{\lambda(2\log_2 N-2)\left(-\left(-\frac{1}{e^{3t\lambda}}+1\right)^2+1\right)}{e^{t\lambda(2\log_2 N-2)}}+6\lambda\left(-\frac{1}{e^{3t\lambda}}+1\right)e^{-2t\lambda\log_2 N-t\lambda}\right)}{-\left(-\frac{1}{e^{3t\lambda}}+1\right)^2+1} \tag{4.65}$$

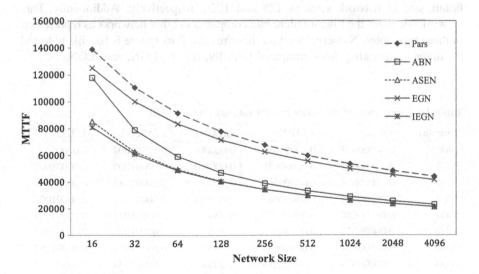

Fig. 4.33 MTTF versus network size

$$\lambda_s(\text{ABN}) = \frac{\lambda\left(\frac{9\log_2(N)-23}{2}\right)\left(\frac{1}{e^{3t\lambda}} + e^{-\frac{15t\lambda}{2}} - e^{-\frac{21t\lambda}{2}}\right) + \frac{15\lambda e^{-\frac{15t\lambda}{2}}}{2} - \frac{21\lambda e^{-\frac{21t\lambda}{2}}}{2} + \frac{3\lambda}{e^{3t\lambda}}}{\frac{1}{e^{3t\lambda}} + e^{-\frac{15t\lambda}{2}} - e^{-\frac{21t\lambda}{2}}}$$

$$(4.66)$$

$$\lambda_s(\text{ASEN}) = \frac{\lambda\left(\frac{9\log_2 N}{2} - 8\right)\left(\frac{1}{e^{3t\lambda}} + e^{-\frac{15t\lambda}{2}} - e^{-\frac{21t\lambda}{2}}\right) + \frac{15\lambda e^{-\frac{15t\lambda}{2}}}{2} - \frac{21\lambda e^{-\frac{21t\lambda}{2}}}{2} + \frac{3\lambda}{e^{3t\lambda}}}{\frac{1}{e^{3t\lambda}} + e^{-\frac{15t\lambda}{2}} - e^{-\frac{21t\lambda}{2}}}$$

$$(4.67)$$

$$\lambda_s(\text{EGN}) = \frac{-\left(-\frac{2\lambda\log_2(N)}{e^{2t\lambda\log_2(N)}}\right)}{e^{(-(2(\log_2(N))))\lambda t}} = 2\lambda\log_2(N)$$

$$(4.68)$$

$$\lambda_s(\text{IEGN}) = \frac{\lambda\left(\frac{9\log_2(N)}{2} - 8\right)\left(\frac{2}{e^{10t\lambda}} - \frac{1}{e^{20t\lambda}} + e^{-\frac{11t\lambda}{2}} + e^{-\frac{51t\lambda}{2}} - 2e^{-\frac{31t\lambda}{2}}\right) + \frac{11\lambda e^{-\frac{11t\lambda}{2}}}{2} + \frac{51\lambda e^{-\frac{51t\lambda}{2}}}{2} - 31\lambda e^{-\frac{31t\lambda}{2}} + \frac{20\lambda}{e^{10t\lambda}} - \frac{20\lambda}{e^{20t\lambda}}}{\frac{2}{e^{10t\lambda}} - \frac{1}{e^{20t\lambda}} + e^{-\frac{11t\lambda}{2}} + e^{-\frac{51t\lambda}{2}} - 2e^{-\frac{31t\lambda}{2}}}$$

$$(4.69)$$

Tables 4.6, 4.7, and 4.8 give the summary of the results of failure rate analysis as a function of time for different network sizes of 16, 128, and 1024, respectively, according to Eqs. (4.65–4.69).

According to Table 4.6, failure rate for different operating times in Pars network is the lowest compared to the other networks. As the results show, it is confirmed that Pars is superior compared to the other networks for the network size of 16, in terms of failure rate. However, we can get useful information by investigation of the failure rate in larger network sizes. Tables 4.7 and 4.8 are provided to indicate the failure rate in network sizes of 128 and 1024, respectively. Additionally, Pars network possesses the lowest failure rate compared to other networks as represented in these two tables. Namely, based on these results, Pars failure is less likely to fail for different operating times compared to ABN, ASEN, EGN, and IEGN.

Table 4.6 Comparison of the failure rate for network size 16×16

Time (h)	λ_s(Pars)	λ_s(ABN)	λ_s(EGN)	λ_s(ASEN)	λ_s(IEGN)
1000	0.00000603	0.00000654	0.000008	0.00001004	0.00001
2000	0.00000607	0.00000659	0.000008	0.00001009	0.00001001
3000	0.00000611	0.00000663	0.000008	0.00001013	0.00001001
4000	0.00000614	0.00000667	0.000008	0.00001017	0.00001002
5000	0.00000618	0.00000672	0.000008	0.00001022	0.00001004
6000	0.00000621	0.00000676	0.000008	0.00001026	0.00001005
7000	0.00000624	0.00000680	0.000008	0.00001030	0.00001007
8000	0.00000628	0.00000684	0.000008	0.00001034	0.00001009

Table 4.7 Comparison of the failure rate for network size 128 × 128

Time (h)	λ_s(Pars)	λ_s(ABN)	λ_s(EGN)	λ_s(ASEN)	λ_s(IEGN)
1000	0.00000603	0.00000654	0.000008	0.00001004	0.00001
2000	0.00000607	0.00000659	0.000008	0.00001009	0.00001001
3000	0.00000611	0.00000663	0.000008	0.00001013	0.00001001
4000	0.00000614	0.00000667	0.000008	0.00001017	0.00001002
5000	0.00000618	0.00000672	0.000008	0.00001022	0.00001004
6000	0.00000621	0.00000676	0.000008	0.00001026	0.00001005
7000	0.00000624	0.00000680	0.000008	0.00001030	0.00001007
8000	0.00000628	0.00000684	0.000008	0.00001034	0.00001009

Table 4.8 Comparison of the failure rate for network size 1024 × 1024

Time (h)	λ_s(Pars)	λ_s(ABN)	λ_s(EGN)	λ_s(ASEN)	λ_s(IEGN)
1000	0.00000603	0.00000654	0.000008	0.00001004	0.00001
2000	0.00000607	0.00000659	0.000008	0.00001009	0.00001001
3000	0.00000611	0.00000663	0.000008	0.00001013	0.00001001
4000	0.00000614	0.00000667	0.000008	0.00001017	0.00001002
5000	0.00000618	0.00000672	0.000008	0.00001022	0.00001004
6000	0.00000621	0.00000676	0.000008	0.00001026	0.00001005
7000	0.00000624	0.00000680	0.000008	0.00001030	0.00001007
8000	0.00000628	0.00000684	0.000008	0.00001034	0.00001009

B. *Fault Tolerance*

Fault tolerance in an interconnection network is very important to operate continuously over a relatively long period of time. Creating redundancy in the number of paths between each pair of source–destination is considered as the basic idea for the fault tolerance of the MIN that leads to gain the ability of use of alternative paths in case of faults.

In this section, number of main and auxiliary paths (i.e., NOMP and NOAP) in a particular basic path between any source–destination pairs will be analyzed in case of fault tolerance of Pars network. The total number of paths between each source–destination pair is computed as follows:

$$\text{Total number of paths} = \text{NOBP} \sum_{n=1}^{\infty} (\text{NOMP}_n + \text{NOAP}_n) \qquad (4.70)$$

Figures 4.34 and 4.35 represent the results of fault-tolerant analysis for the Pars network and other networks, such as ABN [9], ASEN [15, 16], EGN [17], and IEGN [10].

Figure 4.34 indicates that more fault-tolerant capability can be achieved by the Pars network compared to ABN, ASEN, and EGN networks. However, more paths

Fig. 4.34 Total number of paths between each source–destination pair

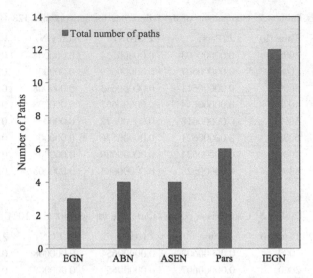

Fig. 4.35 Number of auxiliary and main paths for different networks

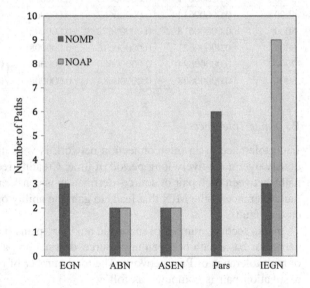

between each source–destination pair can be provided by the IEGN compared to Pars network.

Figure 4.34 shows impressive results, but the types of involved paths cannot be indicated by it. The contribution of each path in fault tolerance is represented in Fig. 4.35. This figure helps to find the path that gives the highest value for each network. Here the question is which of these networks will be performed better and we will try to answer it.

The length of the communications path between every source to its corresponding destination is considered as path length. Multiple paths with different path length are possible in the network. These path lengths can be measured in distance or by counting the number of intermediate switches. Path lengths have the following relations:

$$LOMP_n > LOAP_n, \quad n \in [1, +\infty)$$
$$LOBP_n > LOBP_{n-1}, \quad n \in [2, +\infty)$$

Therefore, according to the fact that the main paths have minimum length between each source–destination pair, the more number of main paths between each source–destination pair, the better situation is for results. Furthermore, more paths are used in Pars compared to the other four networks and in which all paths are of the main types leading to use it rather than the rest.

Determination of the amount of network improvements is defined here in a simple but effective function. Desirability function (DF) is defined as in Eq. (4.71).

$$DF = \frac{Number\ of\ MPs}{Total\ number\ of\ paths} \tag{4.71}$$

With respect to (4.70) and (4.71), we have:

$$DF = \frac{Number\ of\ MPs}{NOBP \sum_{n=1}^{\infty} (NOMP_n + NOAP_n)} \tag{4.72}$$

By using the given source and destination in network, we can obtain this function. Whatever the function of DF be closer to 1, the status of network will be better and reasonable fault-tolerant characteristic will be achieved by network with minimum delay. For networks such as Pars network in which more than one basic path is used, two levels can be considered for the DF: network level and basic path.

Figure 4.36 represented the results of the desirability for the networks of EGN, ABN, ASEN, IEGN, and Pars. Therefore, from this figure we can say that both the Pars and the EGN network are better compare to other three networks in terms of

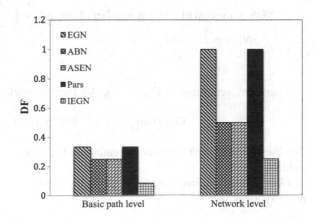

Fig. 4.36 Desirability function for each network

desirability. Here it is considered that despite the same requirements in terms of desirability for both networks of Pars and EGN, the Pars works better than EGN because it has more paths (Fig. 4.34).

C. *Cost*

The most important thing that is considered in MIN design is its cost; as a result, it will not be so useful if its high performance comes at the expense of high cost practically. The cost of a MIN can be estimated by calculation the switch complexity with an assumption that the more number of gates involved, the more cost of a switch. The number of gates involved is in turn roughly proportional to the number of cross points within a switch. For example, a 2×2 switch has four units of hardware cost, whereas a 3×3 switch has nine units. For the multiplexers and demultiplexers, we roughly assume that each of $m \times 1$ multiplexers or $1 \times m$ demultiplexers has m units of cost [9, 17, 19, 47–50].

Based on the above discussion, the following equation is used to calculate the cost of MINs:

$$(\text{NOM} \times \text{the cost of a MUX}) + (\text{NOD} \times \text{the cost of a DMUX}) \\ + ((\text{NOS} \times \text{the cost of a SE}) \times \text{number of stages}) \tag{4.73}$$

where NOM, NOD, and NOS are the number of MUXs, DMUXs, and switches, respectively. We should consider that there are many networks include switches of different sizes. Therefore, calculation of value of $((\text{NOS} \times \text{the cost of a SE}) \times \text{number of stages})$ for each switch must be done separately. For instance, consider the Pars network, we have:

$$(\text{NOM} \times \text{the cost of a MUX}) = \frac{3N}{2} \times 4 = 6N \tag{4.74}$$

$$(\text{NOD} \times \text{the cost of a DMUX}) = \frac{3N}{2} \times 2 = 3N \tag{4.75}$$

$$((\text{NOS} \times \text{the cost of a SE}) \times \text{number of stages}) = \left(\frac{3N}{4} \times 4\right)((\log_2 N) - 1) \\ = 3N((\log_2 N) - 1) \tag{4.76}$$

Therefore, according to Eq. (4.73), the Pars network's cost is given by:

$$\text{Cost(Pars)} = 9N + 3N((\log_2 N) - 1) \tag{4.77}$$

Likewise, the cost function is calculated for each network as it is given in Table 4.9.

Given the cost functions, cost curves are depicted in Fig. 4.37 for different network sizes.

Table 4.9 Cost functions for each network

EGN	ABN	ASEN	Pars	IEGN
$6N + 3N((\log_2 N) - 1)$	$8N + \frac{9}{2}N((\log_2 N) - 3)$	$6N + \frac{9}{2}N((\log_2 N) - 2)$	$9N + 3N((\log_2 N) - 1)$	$6N + \frac{27}{4}N((\log_2 N) - 1)$

Figure 4.37 shows almost the same cost for small sizes in case of all five networks (from 8 to 128), but by increasing in the network size (256 or more), the cost differences become more apparent. The network size in case of Pars and EGN networks is reversely proportional to the cost in comparison with the other networks. Furthermore, the cost of EGN is slightly less than Pars. However, in case of Pars network, there will be greater number of paths between each source–destination pair compared to EGN. As a result, we can get better fault tolerance at the reasonable cost by means of the Pars network.

Here, another concept called as "cost per unit" will be introduced. Cost per unit can be calculated as:

$$\text{Cost per unit} = \frac{\text{Cost}}{\text{Total number of paths}} \qquad (4.78)$$

This parameter indicates how much we are spending per unit of performance. Figure 4.38 shows cost per unit curves for different network sizes. Figure 4.37 indicates almost the same cost per unit results for Pars and IEGN networks in comparison with other three networks. The Pars network and IEGN can get the cost per unit of performance by less spending compared to other three networks and whatever the size of the network increases, it is better to utilize Pars.

Further information can be provided by examining Fig. 4.37; since EGN provides a lower number of paths between each source–destination pair, despite EGN has a lower cost than the other four networks, its cost per unit parameter has quite different situation.

The next parameter that has been emphasized in the most reported works [1, 6, 9, 17, 19] is the cost-effectiveness metric. It is usually considered as a measure of the efficiency of a MIN in terms of cost. Therefore, this parameter will be discussed to a

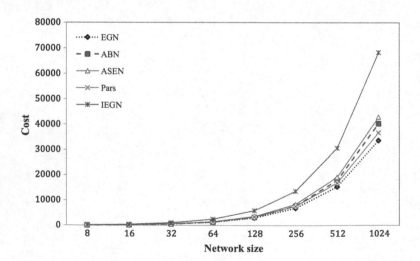

Fig. 4.37 Cost versus network size

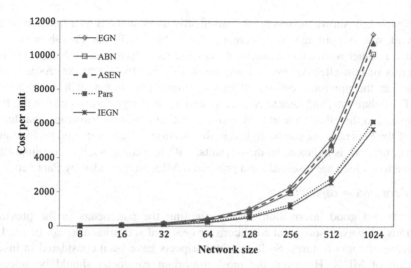

Fig. 4.38 Cost per unit versus network size

comprehensive review of the hardware costs of the networks. The cost-effectiveness (CE) parameter is given by Eq. (4.79) [1, 6, 9, 17, 19].

$$\text{CE} = \frac{\text{Mean time to failure}}{\text{Cost}} = \frac{\int_0^\infty R(t)\mathrm{d}t}{\text{Cost}} \tag{4.79}$$

In this section, like [9, 11, 17, 19, 20], the exponential distribution is intended to describe the times to failure of the switch components. In addition, λ can be considered 10^{-6} per hour [9, 19, 24]. Therefore, according to Eq. (4.79), we have:

$$\text{CE(Pars)} = \frac{\int_0^\infty \left(e^{-(2(\log_2 N)-2)\lambda t}\left(1 - (1 - e^{-3\lambda t})^2 \right) \right)\mathrm{d}t}{9N + 3N((\log_2 N) - 1)} \tag{4.80}$$

$$\text{CE(ABN)} = \frac{\int_0^\infty \left(e^{-\left(\frac{9}{2}(\log_2 N)-\frac{23}{2}\right)\lambda t}\left(1 - \left((1 - e^{-3\lambda t})\left(1 - e^{-\frac{15}{2}\lambda t}\right) \right) \right) \right)\mathrm{d}t}{8N + \frac{9}{2}N((\log_2 N) - 3)} \tag{4.81}$$

$$\text{CE(ASEN)} = \frac{\int_0^\infty \left(e^{-\left(\frac{9}{2}(\log_2 N)-8\right)\lambda t}\left(1 - \left((1 - e^{-3\lambda t})\left(1 - e^{-\frac{15}{2}\lambda t}\right) \right) \right) \right)\mathrm{d}t}{6N + \frac{9}{2}N((\log_2 N) - 2)} \tag{4.82}$$

$$\text{CE(EGN)} = \frac{\int_0^\infty \left(e^{-(2(\log_2 N))\lambda t} \right)\mathrm{d}t}{6N + 3N((\log_2 N) - 1)} \tag{4.83}$$

CE(IEGN)

$$= \frac{\int_0^\infty \left(\left(e^{-\left(\frac{9}{4}(\log_2 N)-8\right)\lambda t} \right)\left(e^{-\frac{9}{2}\lambda t}\left(1 - \left(1 - e^{-\frac{9}{2}\lambda t}\right)\left(1 - \left(1 - \left(1 - e^{-\frac{9}{2}\lambda t}\right)(1 - e^{-10\lambda t})\right)\right)\right) \right) + \left(1 - e^{-\frac{9}{2}\lambda t}\right)\left(1 - \left(1 - e^{-\frac{9}{2}\lambda t}\right)(1 - e^{-10\lambda t})\right) \right)\mathrm{d}t}{6N + 27/4N((\log_2 N) - 1)} \tag{4.84}$$

Figure 4.39 shows the results of cost-effectiveness analysis as a function of the network size. According to this figure, the networks of EGN and Pars show the best results. In other words, the networks of Pars and EGN have almost the same results in terms of cost-effectiveness in comparison to other two networks. As it is discussed in the previous sections, in case of many parameters such as reliability, MTTF, failure rate, and fault tolerance, we can get better performance by using Pars compared to the other networks. Moreover, getting desirable hardware cost due to use of the Pars network can be argued in this section. In fact, the good performance of Pars network is indicated by these results, with less costs. So, the most important parameters which are essential for a prospered MIN are provided by Pars network.

D. *Permutation capability*

We can get good information by investigating the parameters in the previous sections. However, some real conditions or near-real scenarios also can be used in studying of some features. So far, various aspects have been considered in investigation of MINs. However, the most important parameter should be selected according to the ability to express the main purpose of evolving these networks. The proper delivery of requests from the sources to the destinations is the main objective of these networks. Therefore, in this subsection, investigation of the parameter of permutation capability can serve two purposes for us: first is to understand how we can verify the previous parameters. Second, the reality of network performance can be shown by this parameter. Different numbers of requests will be considered to measure the permutations capability, and the success rate of the network in case of performing these requests will be examined. In fact, this parameter tells how many requests appeared at source side has successfully got matured (e.g., how many requests have successfully reached their destinations). In what follows, we will get more acquainted with the following quantities.

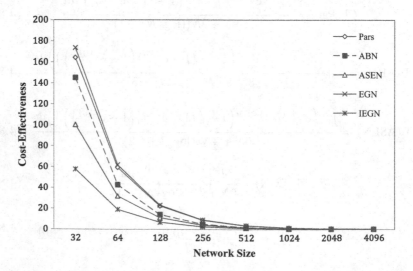

Fig. 4.39 Cost-effectiveness versus network size

(1) *Arrival rate*

Arrival rate is the number of requests arrived successfully.

(2) *The mean time to arrival (MTTA)*

MTTA is the average time that a specific request takes to reach its destination. To calculate MTTA, we use the following equation:

$$\text{MTTA} = \frac{\text{Total time}}{\text{Arrival rate}} \qquad (4.85)$$

The sum of the individual response times to requests is considered to calculate the total time.

(3) *Response time*

The time takes to respond to a series of requests is called as response time. In fact, this parameter defines the time at the network level that can be defined as in Eq. (4.86).

$$\text{Response time} = \text{Max}(T(R_1), T(R_2), \ldots, T(R_n)) \qquad (4.86)$$

where $T(R_i)$ refers to the response time to i-th request.

(4) *Efficiency*

We define efficiency as in Eq. (4.87).

$$\text{Efficiency} = \frac{\text{arrival rate}}{\text{Traffic rate}} \qquad (4.87)$$

Here, we have the following assumptions:

1. Size 8 is considered for networks.
2. Each group of requests is simultaneously applied to the network.
3. In this study, delivery time is computed regarding to switching delay across the path plus the time for request issuing by the source.
4. Like [23], in this research switching delay is assumed to be 0.1 ms.
5. To our best knowledge, none of the reported works has considered the request issue. Therefore, for the first time, we take this parameter into account that is 0.1 ms.

The smallest unit of communication known as packet contains the packet header in which there is the destination address and sequencing information. For the topologies in which packets may have to traverse some intermediate switching elements, the routing algorithm determines the path selected by a packet to reach its destination. The next channel to be used is indicated by means of the routing algorithm, at each intermediate switch. We can select that channel among a set of possible choices. The packet is blocked and cannot advance due to being busy all the candidate channels.

In general, since the communication resources such as switching elements and links are limited and shared in an interconnection network, contentions and deadlocks arise. When several packets flows concurrently request access to the same output port from different input ports, congestion happens inside a switch and its origins contention [51]. In these cases, only one packet can cross at a given moment (this packet is selected randomly in this study), while the other packets contending for the output port should have chosen other paths to reach their destination. If all the paths are unavailable, then the packet is blocked.

When the packets cannot move toward their destination due to the business of input port requested, deadlock occurs. In case all the candidate input ports are busy, the packets involved in a deadlocked configuration will be blocked forever. It should be considered that a packet may be permanently blocked in the network because the destination node does not consume it. This type of deadlock is caused by the application and is beyond the scope of this paper. In this subsection, the assumption is that packets are always consumed by the destination node. Consequently, a set of packets is blocked forever in a deadlocked configuration. Each packet asks for resources held by other packet(s) and at the same time holds resources requested by other packet(s). In other words, a message is delivered between terminals by making several hops across the shared links and switch nodes from its source terminal to destination terminal. Therefore, the contention and deadlock problems originate from the fact that communication resources such as switching elements and links are limited and shared in an interconnection network. As aforesaid, in this study the assumption is that the main way to deal with contention or deadlock is utilizing alternative paths.

In Fig. 4.40, the results of arrival rate analysis are shown.

As we see in Fig. 4.40, each of the four networks can equally serve in a low traffic rate (up to 33%). However, ABN presents weak performance at high traffic rates compared to other three networks. Besides, both ASEN and EGN show the same results in all traffic rate scenarios. They present better performance in both traffic rates of 66.66 and 83.33% compared to Pars network. But in the traffic rate of 100%, Pars network outperforms EGN and ASEN in terms of arrival rate.

In order to understand which networks perform better, the slope of each curve should be considered. Evidently, both EGN and ASEN attain better arrival rate at traffic rate of 66.66%, but afterward, two aforementioned networks do not tend to be improved. However, the slope of the Pars network is constantly increasing; consequently, the maximum arrival rate for the Pars network is greater than the other networks. In fact, according to the results, even in high traffic rates trend of Pars network performance is continuously improved.

MTTA analysis results are presented in Fig. 4.41.

MTTA is the average time required for each request to reach its corresponding destination. Unlike Figs. 4.40 and 4.41 reveals that ABN outperforms other three networks in terms of MTTA, and this superiority is due to the fewer number of stages in ABN compared to the rest. This causes ABN not to have a good performance as depicted in Fig. 4.40. In fact, a network is appropriate once it improves

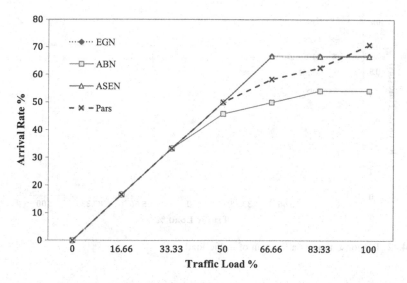

Fig. 4.40 Arrival rate as a function of traffic load

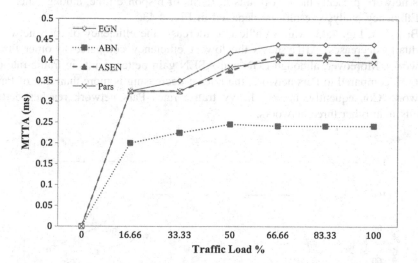

Fig. 4.41 Mean time to arrival (MTTA) versus traffic load

most of the performance parameters. According to Fig. 4.41, after ABN, the Pars network presents the best results in terms of MTTF.

In Fig. 4.42, response time analysis results are presented.

In Fig. 4.42, we face a situation similar to Fig. 4.41. That is, although ABN presents better results in terms of response time compared to other three networks, this does not imply superiority of ABN when considering its poor performance in terms of arrival rate, cost, and reliability.

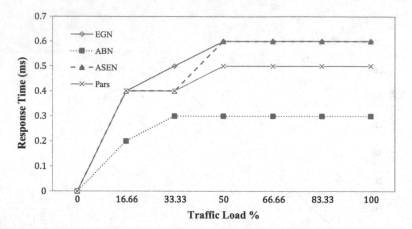

Fig. 4.42 Response time as a function of traffic load

Pars network not only results in acceptable performance in terms of cost and arrival rate, but also has a good performance in terms of response time. After ABN, Pars network presents the best results in terms of response time, among others.

Efficiency analysis results are depicted in Fig. 4.43.

Based on Fig. 4.43, with a traffic rate increase, the efficiency of each network gradually decreases. ABN shows the lowest efficiency compared to other three networks. Moreover, although EGN and ASEN gain better results in traffic rate of 66.66% compared to Pars network, the rate of decreasing is more than that of Pars network. Consequently, in very heavy traffic rates Pars network reaches better results than other three networks.

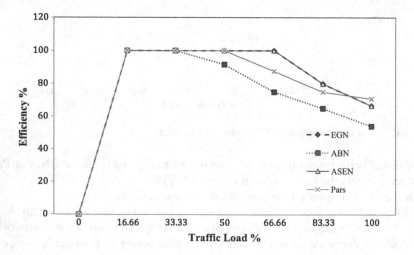

Fig. 4.43 Efficiency versus traffic load

We can generally summarize the results of the analysis of terminal reliability, fault tolerance, cost, and permutation capability as follows:

1. Pars network shows higher terminal reliability, mean time to failure (MTTF), and failure rate than other four networks.
2. Pars network supplies larger number of paths between each source–destination pair compared to three popular networks of ABN, ASEN, and EGN. As a result, Pars networks get better fault-tolerant characteristics.
3. Pars network is also more acceptable in terms of cost; it leads to better results cost-wise compared with three networks, such as ABN, ASEN, IEGN. However, it is slightly more expensive than the EGN network. Pars network also obtains impressive results in terms of "cost per unit" and "cost-effective" parameters compared to earlier representative networks.
4. Permutation capability analysis demonstrates that Pars network has acceptable performance, even in heavy traffic rates scenarios. Results depict that the most distinguished feature of Pars network among other networks is lower slope decreasing. It also has a reasonable response time.

We have introduced a fault-tolerant MIN in this section, named Pars network, which is a low-cost network due to its small-size switching elements (i.e., 2×2). It also provides multiple paths between each pair of source–destination, which makes the network be robust against faults and reduces blocking problem. The extensive performance analysis results demonstrate that Pars network outperforms known regular MINs, namely ABN, ASEN, EGN, and IEGN in terms of cost, fault tolerance, terminal reliability, MTTF, and permutation capability. Moreover, self-routing mechanism in Pars simplifies it, at the same time effective. According to the results, Pars network is the perfect candidate for multiprocessor systems. However, it is believed that Pars network can present even better performance when combined with other modern design approaches. For example, recent researches show that multilayer MINs provide better performance which should be deeply investigated to reach the most appropriate designs.

4.4 Improving the Reliability of the Benes Network

Rearrangeable non-blocking network is one of the most important types of fault-tolerant topologies in multistage interconnection networks. These networks are able to provide all the required connections in a permutation, without blocking. However, to accomplish this task, it may be necessary to reorganize some of the current connections in these networks. Hence, rearrangeable non-blocking MINs are known as the main options to solve the blocking problem.

One main advantage of rearrangeable non-blocking networks compared to other ones such as Clos (i.e., a non-blocking network) is lower network complexity and consequently lower hardware cost. One well-known topology in this network category is Benes network, which has been examined in many researches.

Reliability is an important aspect of performance to be considered in the system assessment. Therefore, improving the reliability and fault tolerance in Benes network can be one of the leading fields of research in the realm of non-blocking networks. Although good works have been conducted to improve the reliability and fault tolerance of blocking networks, less attention has been paid to this performance aspect in Benes network (Benes network is a rearrangeable non-blocking network). Therefore, in this section we focus on analyzing the reliability of Benes network as well as offering effective methods to improve the reliability and fault tolerance of this network.

Studying previous works reveals that three main methods can be considered to improve reliability in MINs: (1) increasing the number of stages, (2) putting multiple networks in parallel, and (3) taking advantage of replicated MINs. To find the best solution to improve the reliability of Benes network, all three of these methods will be investigated in this section.

4.4.1 Different Methods for Improving Reliability in Benes Network

Three important methods to enhance reliability in MINs will be studied about Benes network in this section. Subsection A is intended to examine the idea of increasing the number of stages in the Benes network. Subsection B is devoted to examining the approach of using several Benes networks in parallel. In addition, exploiting the replicated MINs on Benes network is presented in subsection C. Each of these approaches has some weaknesses and strengths. However, an approach can be a better choice, if it gets more improvement than other approaches in terms of reliability. After reviewing various approaches in this section, their performance will be calculated in Sect. 4.4.2. Finally, based on the analysis results, the most efficient approach for enhancing reliability in Benes network will be specified.

First of all, it is better to know the Benes network structure. A Benes network can be considered as a combination of a Baseline network (i.e., a blocking MIN) and an inverted Baseline network (i.e., a blocking MIN) in which the middle stages overlap. A Benes network of size $N \times N$ consists of $(2(\log_2 N) - 1)$ stages. Also, each stage is made up of $\left(\frac{N}{2}\right)$ switches of size 2×2. The network complexity of an $N \times N$ Benes network is $\left[\frac{N}{2}(2(\log_2 N) - 1)\right]$. A Benes network of size 8×8 is shown in Fig. 4.44.

A. Extra-stage Benes network

Here, the method of "increasing the number of stages," is used in Benes network. Also, the obtained topology is named extra-stage Benes network (EBN). Here, the main methodology is to increase the number of routes from any source to any specific destination that results in improved reliability. In fact, an $N \times N$ EBN is an $N \times N$ Benes network, with the difference that a stage is added to it. Figure 4.45

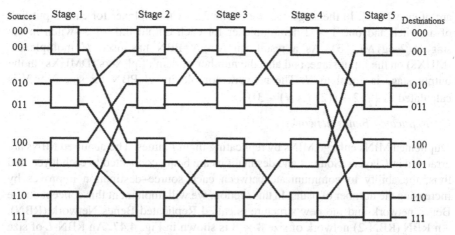

Fig. 4.44 A Benes network of size 8 × 8

illustrates an EBN of size 8 × 8. An $N \times N$ EBN is made up of $(2(\log_2 N))$ stages. Also, the number of 2 × 2 switches in each of these stages is $\left(\frac{N}{2}\right)$. Therefore, the network complexity of this MIN will be equal to $\left[\frac{N}{2}(2(\log_2 N))\right]$.

B. Parallel Benes network

Using multiple networks in parallel is another idea to improve the reliability of MINs that is investigated on Benes network in this subsection. The obtained topology is named parallel Benes network (PBN). Here, the main idea is to generate redundancy in the number of paths using multiple separate networks connected within a larger network. This strategy can increase the number of routes from any source to any destination in the network. In PBN's network structure, there are two subnetworks that each one is a Benes network. A 16 × 16 PBN is presented in Fig. 4.46. An $N \times N$ PBN has $((2 \log_2 N) - 3)$ stages of $\left(\frac{N}{2}\right)$ switches. All switches

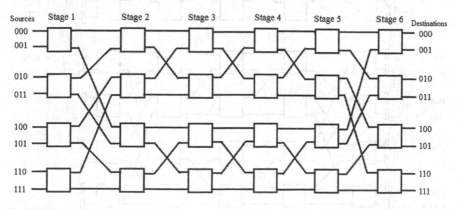

Fig. 4.45 A 8 × 8 EBN network

are of size 2×2. In the stage 1, there is one 2×1 multiplexer for each input link of a switch and one 1×2 demultiplexer for each output link of a switch in the stage $((2 \log_2 N) - 3)$. As a result, an $N \times N$ PBN has used N multiplexers (MUXs) on the input stage, and also the number of demultiplexers (DMUXs) in the output stage is equal to N. The network complexity of PBN for size $N \times N$ is calculated as $\left[\frac{N}{2} (2 + ((2 \log_2 N) - 3)) \right]$.

C. *Replicated Benes networks*

Duplicated MINs enlarge MINs by replicating them L times. The acquired MINs are arranged in L layers. Sources and destinations are being connected to each layer, and thus, the ability to communicate between each source–destination improves by increasing the number of paths. In this section, we will implement this concept on the Benes network and the new structure is called Replicated Benes Network (RBN). An RBN (RBN-2) network of size 8×8 is shown in Fig. 4.47. An RBN-L of size $N \times N$ is designed by replicating Benes network L times. There are L independent Benes networks, such that a connection path that flows within a source to any destination can be seen via any one of the networks. All the L subnetworks are of identical type. Each source and destination is connected to all the L subnetworks. An RBN-L composed of $((2 \log_2 N) - 1)$ stages and $\left[\left(\frac{N}{2} (2 (\log_2 N) - 1) \right) \right]$ switches.

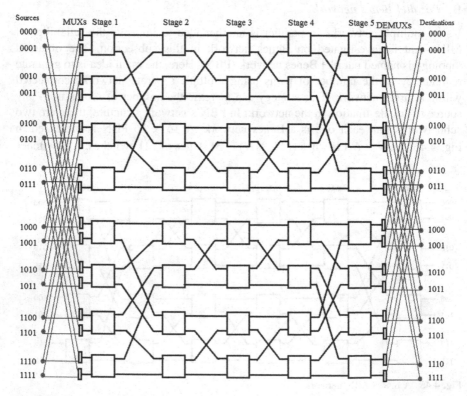

Fig. 4.46 A 16×16 PBN network

The network complexity of an $N \times N$ RBN-L with $L = 2$ is equal to $(N(2(\log_2 N) - 1))$. It needs to be reminded that we limit our learning process to RBN-2 (RBN), in which the number of layers L is equal to 2.

4.4.2   Reliability Analysis of EBN, PBN, and RBN Networks

We will examine the reliability from two important aspects, terminal and broadcast in this subsection. The probability of at least one non-faulty path from any given source–destination pair is defined as terminal reliability. In addition, the probability of at least non-faulty path from one particular source to all network destinations is referred as broadcast reliability. Also, reliability analysis can be done in two forms: time-independent or time-dependent. Usually, one of them has been employed in the previous works. Nevertheless, in this subsection, both scenarios will be examined in order to carry out thorough examination.

The other prominent fact is related to the cost in any attempt to improving the reliability. Approaches to enhance reliability are often limited by increased hardware costs in the domain of MINs. Therefore, an approach that has a high cost to the network may not be affordable in practice. In previous works [1, 6, 9, 52, 53], a metric has been considered as a parameter in which the results are effective in relation to its cost (mean time to failure (MTTF)/cost ratio), which is an important metric that can be utilized in analysis of the MINs efficiency from a cost point of view. As a result, we will study this metric to a suitable analysis of the networks' hardware costs.

In fact, based on the above discussions, in this subsection we will analyze the reliability in terms of the following parameters:

- Terminal reliability (time-independent reliability)
- Broadcast reliability (time-independent reliability)
- Terminal time-dependent reliability
- Broadcast time-dependent reliability
- Cost-effectiveness.

Based on detailed analysis of metrics mentioned above, it can be said that the reliability examination carried out in this subsection results in valid outcomes.

Time-independent terminal reliability, time-independent broadcast reliability, time-dependent terminal reliability, time-dependent broadcast reliability, and cost-effectiveness parameter will be analyzed in subsection A through F in order.

A.   Terminal reliability (time-independent analysis)

We have used the switch fault model for reliability analysis of MINs in this section. In this model, it is presumed that any switching component (i.e., switching elements, multiplexers, and demultiplexers) are prone to failure. Also, it is assumed that the reliability of a 2×2 switch (SE$_2$) is equal to r.

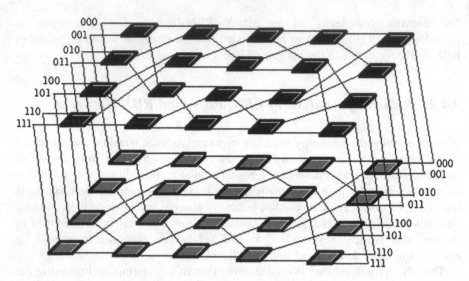

Fig. 4.47 3-D view of a RBN network of size 8 × 8

Based on Benes network's topology (from Fig. 4.44), terminal reliability RBD of 8 × 8 Benes network is presented in Fig. 4.48. Based on this diagram, terminal reliability of 8 × 8 Benes network, denoted $R_t(8 \times 8 \text{ Benes})$, is obtained from Eq. (4.88):

$$R_t(8 \times 8 \text{ Benes}) = r^2 \left[1 - \left(1 - \left(r^2 \left(1 - (1-r)^2 \right) \right) \right)^2 \right] \qquad (4.88)$$

In addition, based on the topology of the Benes network, terminal reliability RBD of 16 × 16 Benes network is illustrated in Fig. 4.49. According to this figure, the terminal reliability of an $N \times N$ Benes network is calculated as below:

Fig. 4.48 Terminal reliability RBD of 8 × 8 Benes network

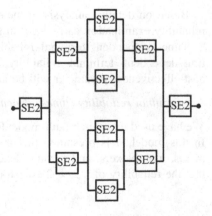

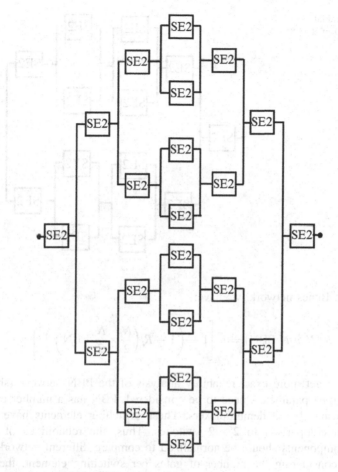

Fig. 4.49 Terminal reliability RBD of 16 × 16 Benes network

$$R_t(N \times N \text{ Benes}) = r^2 \left[1 - \left(1 - R_t \left(\frac{N}{2} \times \frac{N}{2} \text{ Benes} \right) \right)^2 \right] \qquad (4.89)$$

Considering the structure of EBN (in Fig. 4.45), the terminal reliability RBD of 8 × 8 EBN is illustrated in Fig. 4.50. Based on following figure, the terminal reliability of 8 × 8 EBN is determined as below:

$$R_t(8 \times 8 \text{ EBN}) = r^2 \left[1 - \left(1 - \left(r^2 \left(1 - (1 - r)^2 \right)^2 \right) \right)^2 \right] \qquad (4.90)$$

Fig. 4.50 Terminal
reliability RBD of 8×8
EBN network

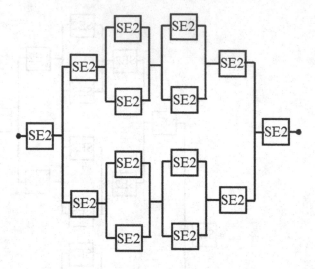

Similar to Benes network, we have:

$$R_t(N \times N \text{ EBN}) = r^2 \left[1 - \left(1 - R_t\left(\frac{N}{2} \times \frac{N}{2} \text{ EBN}\right)\right)^2 \right] \qquad (4.91)$$

In order to get more exact reliability analysis of the PBN network (shown in Fig. 4.46), some parameters need to be considered. PBN has a number of 2×1 multiplexers and 1×2 demultiplexers. These switching elements have various reliability in comparison to 2×2 switches. Thus, the reliabilities of various switching components should be normalized to compare different networks properly. Here, considering the number of gates per switching element, the switch reliability can be calculated based on r. It is presumed that the hardware complexity of an element is straightly corresponding to the gate counts [6, 9, 20, 52].

RBD of 16×16 PBN for terminal reliability is illustrated in Fig. 4.51. In Fig. 4.51, switching elements of size 2×2 are shown as SE_2, and 2×1 multiplexers and 1×2 demultiplexers are illustrated as MUX_2 and DEMUX_2, in order. Considering the terminal reliability RBD of 16×16 PBN (shown in Fig. 4.51), the terminal reliability of 16×16 PBN is given by Eq. (4.92).

$$R_t(16 \times 16 \text{ PBN}) = 1 - \left(1 - \left(r^3 \left[1 - \left(1 - \left(r^2\left(1 - (1 - r)^2\right)\right)\right)^2 \right]\right)\right)^2$$

$$(4.92)$$

Fig. 4.51 Terminal reliability RBD of 16 × 16 PBN network

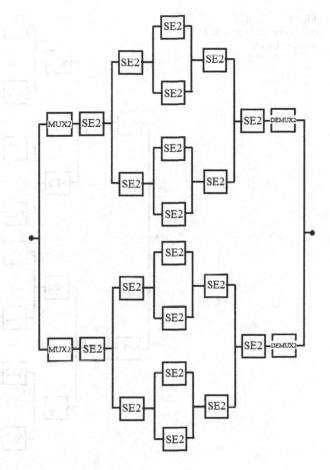

In addition, to $N \times N$ PBN, we have:

$$R_t(N \times N \text{ PBN}) = 1 - \left(1 - r\left[R_t\left(\frac{N}{2} \times \frac{N}{2} \text{ Benes} \right) \right] \right)^2 \qquad (4.93)$$

According to RBN network topology (illustrated in Fig. 4.47), the terminal reliability RBD is obtained in Fig. 4.52. Reliability for this figure is obtained from the equation below:

$$R_t(8 \times 8 \text{ RBN}) = 1 - \left(1 - \left(r^2\left[1 - \left(1 - \left(r^2\left(1 - (1 - r)^2 \right) \right) \right)^2 \right] \right) \right)^2 \qquad (4.94)$$

Fig. 4.52 Terminal
reliability RBD of 8 × 8
RBN network

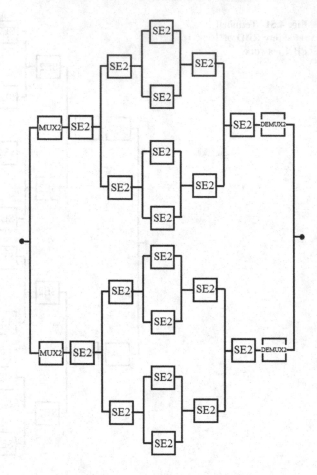

Furthermore, for RBN of size $N \times N$, we have:

$$R_t(N \times N \, \text{RBN}) = 1 - (1 - (R_t(N \times N \, \text{Benes})))^2 \qquad (4.95)$$

Figures 4.53, 4.54, and 4.55 show terminal reliability outcomes of Benes network, EBN, PBN, and RBN for various network sizes. In this analysis, three values have been considered for switch reliability (r): 0.9 (low switch reliability), 0.95 (middle switch reliability), and 0.99 (high switch reliability).

All three Figs. 4.53, 4.54, and 4.55 demonstrate almost the similar results and that is Benes and EBN networks have the least terminal reliability compared to the other networks. These two networks are like each other in case of terminal reliability as well. However, Benes network is rather more suitable for some network sizes than the EBN. It can be resulted that the method of increasing the number of stages to the Benes network does not have any advantages on reliability. The major reason is that increasing network depth (number of stages) of EBN in comparison to Benes network causes the network structure to become more complicated.

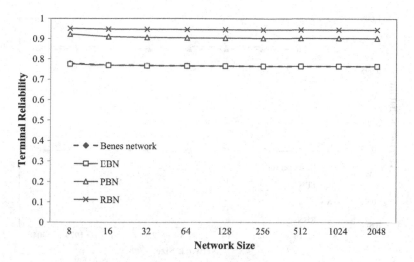

Fig. 4.53 Terminal reliability versus network size in low switch reliability ($r = 0.9$)

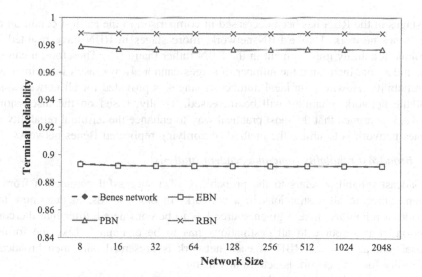

Fig. 4.54 Terminal reliability versus network size in middle switch reliability ($r = 0.95$)

Contrarily, according to previous studies, adding network sophistication can result in decreasing the reliability as the number of stages increases [7, 19].

In addition, these results demonstrate that two networks of PBN and RBN own the best results. Thus, much higher level of reliability can be attained in comparison to the other two networks. Nonetheless, as the figures shows, RBN attains higher terminal reliability compared to PBN. The explanation behind this is that the PBN network does not have as many stages as the RBN. On the other hand, the number

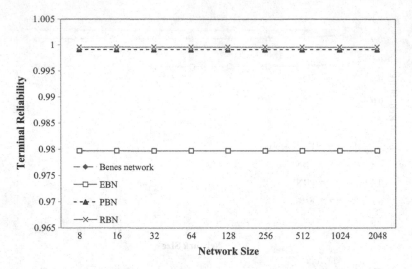

Fig. 4.55 Terminal reliability versus network size in high switch reliability ($r = 0.99$)

of stages in the RBN has been decreased in comparison to the earliest condition of the Benes network. Unlike EBN network, more stages in RBN have resulted in terminal reliability improvement in the RBN rather than PBN. Therefore, it can be concluded that increasing the number of stages cannot always result in an increase in reliability. However, an ideal number of stages is provided for all networks; so that the network reliability will be increased. Totally, based on the discussions above, it is argued that the most practical way to enhance the terminal reliability of Benes network is to utilize the method of applying replicated Benes networks.

B. *Broadcast reliability (time-independent analysis)*

Broadcast reliability refers to the probability of a successful connection from a given source to all destinations in a network. In other words, to determine the broadcast reliability, first, a given source has to be considered, after that the connection of this source to all destinations has to be examined. Like the former subsection, at first, the RBD of each network is presented, and then broadcast reliability for the network is determined using it.

Considering Benes network's topology (visible in Fig. 4.44), its broadcast reliability RBD can be illustrated as in Fig. 4.56. Based on this figure, the broadcast reliability of 8×8 Benes network is obtained as follows:

$$R_b(8 \times 8 \text{ Benes}) = r^5 \left[1 - \left(1 - \left(r^3 \left(1 - (1 - r)^2 \right) \right) \right)^2 \right] \qquad (4.96)$$

Furthermore, based on the diagram shown in Fig. 4.57, which is related to the Benes network of size 16×16, we can argue that the network has a behavior that its broadcast reliability for size $N \times N$ is calculated using Eq. (4.97).

Fig. 4.56 Broadcast reliability RBD of 8 × 8 Benes network

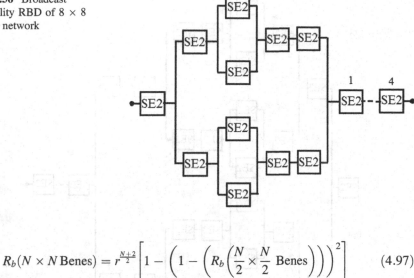

$$R_b(N \times N \text{ Benes}) = r^{\frac{N+2}{2}} \left[1 - \left(1 - \left(R_b \left(\frac{N}{2} \times \frac{N}{2} \text{ Benes} \right) \right) \right)^2 \right] \qquad (4.97)$$

Broadcast reliability RBD for 8 × 8 EBN network is illustrated in Fig. 4.58. Considering the diagram, broadcast reliability of 8 × 8 EBN is calculated as shown below:

$$R_b(8 \times 8 \text{ EBN}) = r^5 \left[1 - \left(1 - \left(r^3 \left(1 - (1 - r)^2 \right)^2 \right) \right)^2 \right] \qquad (4.98)$$

Similarly, the following equation can be used to calculate the broadcast reliability of the EBN in larger network sizes.

$$R_b(N \times N \text{ EBN}) = r^{\frac{N+2}{2}} \left[1 - \left(1 - \left(R_b \left(\frac{N}{2} \times \frac{N}{2} \text{ EBN} \right) \right) \right)^2 \right] \qquad (4.99)$$

Based on Figs. 4.59 and 4.60, broadcast reliability of 16 × 16 PBN and 8 × 8 RBN is given by (4.100) and (4.101):

$$R_b(16 \times 16 \text{ PBN}) = 1 - \left(1 - \left(r^{\frac{19}{2}} \left[1 - \left(1 - \left(r^3 \left(1 - (1 - r)^2 \right) \right) \right)^2 \right] \right) \right)^2 \qquad (4.100)$$

$$R_b(8 \times 8 \text{ RBN}) = 1 - \left(1 - \left(r^5 \left[1 - \left(1 - \left(r^3 \left(1 - (1 - r)^2 \right) \right) \right)^2 \right] \right) \right)^2 \qquad (4.101)$$

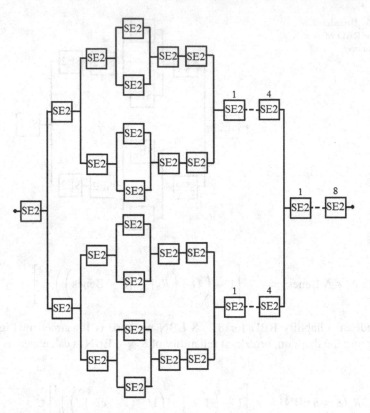

Fig. 4.57 Broadcast reliability RBD of 16 × 16 Benes network

Fig. 4.58 Broadcast
reliability RBD of 8 × 8
EBN network

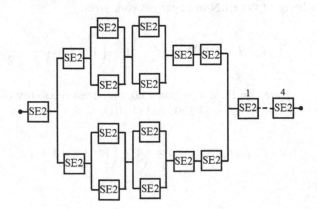

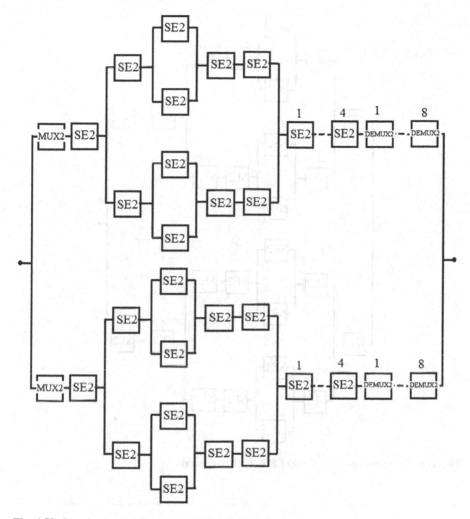

Fig. 4.59 Broadcast reliability RBD of 16 × 16 PBN network

Moreover, for network size of $N \times N$, it is calculated as below:

$$R_b(N \times N \, \text{PBN}) = 1 - \left(1 - \left(r^{\frac{N+2}{4}}\left(R_b\left(\frac{N}{2} \times \frac{N}{2} \, \text{Benes}\right)\right)\right)\right)^2 \quad (4.102)$$

$$R_b(N \times N \, \text{RBN}) = 1 - (1 - (R_b(N \times N \, \text{Benes})))^2 \quad (4.103)$$

Figures 4.61 and 4.62 show broadcast reliability thorough examination outcomes for various network sizes and switch reliabilities.

Figures 4.61, 4.62, and 4.63 show the most and the least broadcast reliability that are PBN and EBN own for various network sizes and switch reliabilities.

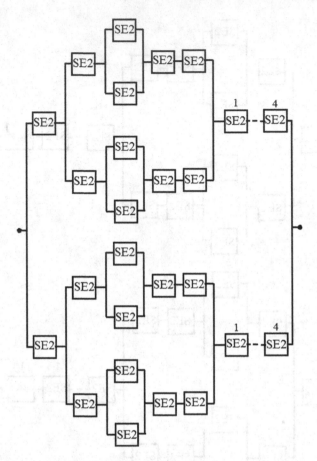

Fig. 4.60 Broadcast reliability RBD of 8 × 8 RBN network

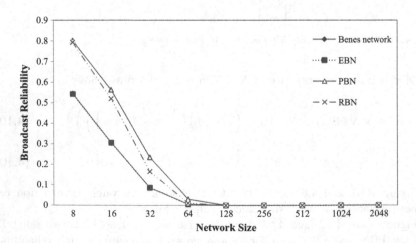

Fig. 4.61 Broadcast reliability versus network size in low switch reliability ($r = 0.9$)

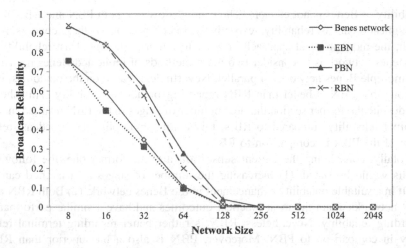

Fig. 4.62 Broadcast reliability versus network size in middle switch reliability ($r = 0.95$)

Broadcast reliability can be decreased as the result of an increase in the network complexity which occurs because of the increased stages. Consequently, EBN achieves the lowest reliability compared to the other networks. Nevertheless, as the network size and switch reliability multiply, the reliability of two networks, Benes and EBN, are more associated with each other so that their performance becomes very close to the switch reliability of 0.99 (Fig. 4.63).

The approach of increasing stages in order to enhance the reliability of Benes is not efficient in any ways. Performance of EBN is not desirable considering the hardware cost. Therefore, the approach of increasing stages in order to enhance the

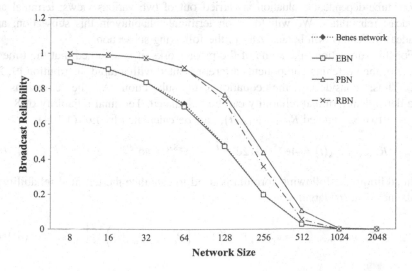

Fig. 4.63 Broadcast reliability versus network size in high switch reliability ($r = 0.99$)

reliability of Benes is not efficient. The results show that both PBN and RBN have the highest broadcast reliability, respectively, in comparison to other networks. As a result, the most practical approach in order to enhance the broadcast reliability of the Benes network is to consider two more methods of replicated Benes networks and multiple Benes networks in parallel. Nevertheless, based on the outcomes, it is obvious that PBN is better than RBN regarding broadcast reliability. This shows that despite the former section that less number of stages in the PBN resulted in less terminal reliability compared to RBN, this feature multiplies the broadcast reliability of the PBN in comparison to RBN.

Totally, considering the current subsection and the former one, the following results would be stated: (1) Increasing the number of stages as a method cannot result in a suitable reliability enhancement in the Benes network. (2) Both RBN and PBN networks attain the most practical outcomes and have a similar performance regarding reliability. Nevertheless, RBN is rather better regarding terminal reliability in comparison to PBN. Moreover, PBN is also a bit superior than RBN regarding broadcast reliability. In fact, utilizing replicated Benes networks and multiple parallel Benes networks together is the most beneficial option. However, which of these approaches can achieve better functionality? To get the answer, the following time-dependent reliability examination is useful.

C. *Terminal reliability (time-dependent analysis)*

All analyses in prior subsections were based on time-independent reliability examination. This kind of reliability thorough examination can present important data, and most of the prior researches were carried out due to this kind of examination [7, 19, 22]. Nevertheless, a time-dependent analysis of a system is also important as well as reliability for a more comprehensive analysis. Therefore, we will also examine the time-dependent analysis in this subsection. Like time-independent reliability examination, time-dependent evaluation is carried out of two various views; terminal and broadcast reliabilities. We will focus on terminal reliability in this subsection, and broadcast reliability will be analyzed in the following subsection.

For this subsection, like some of the prior works, it is assumed that the time to failure of the switching components are represented with a rapid distribution [9, 20, 52]. Thus, considering the equations in subsection A, the equations of time-dependent terminal reliability can be calculated. Terminal reliability of 8×8 Benes network, denoted $R_{t-8 \times 8 \text{ Benes}}(t)$, can be calculated by Eq. (4.104).

$$R_{t-8 \times 8 \text{ Benes}}(t) = 4\mathrm{e}^{-5\lambda t} - 2\mathrm{e}^{-6\lambda t} - 4\mathrm{e}^{-8\lambda t} + 4\mathrm{e}^{-9\lambda t} - \mathrm{e}^{-10\lambda t} \qquad (4.104)$$

In addition, the following formula is used to calculate the terminal reliability of Benes network for larger sizes:

$$R_{t-N \times N \text{ Benes}}(t) = \mathrm{e}^{-2\lambda t}\left[1 - \left(1 - \left(R_{t-\frac{N}{2} \times \frac{N}{2} \text{ Benes}}(t)\right)\right)^2\right] \qquad (4.105)$$

Similarly, for the networks of EBN, PBN, and RBN, we have:

$$R_{t-8\times 8\ \text{EBN}}(t) = 8e^{-6\lambda t} - 8e^{-7\lambda t} + 2e^{-8\lambda t} - 16e^{-10\lambda t}$$
$$+ 32e^{-11\lambda t} - 24e^{-12\lambda t} + 8e^{-13\lambda t} - e^{-14\lambda t} \tag{4.106}$$

$$R_{t-N\times \text{NEBN}}(t) = e^{-2\lambda t}\left[1 - \left(1 - \left(R_{t-\frac{N}{2}\times\frac{N}{2}\text{EBN}}(t)\right)\right)^2\right] \tag{4.107}$$

$$R_{t-16\times 16\ \text{PBN}}(t) = 8e^{-6\lambda t} - 4e^{-7\lambda t} - 8e^{-9\lambda t} + 8e^{-10\lambda t} - 2e^{-11\lambda t}$$
$$- 16e^{-12\lambda t} + 16e^{-13\lambda t} + 32e^{-15\lambda t} - 48e^{-16\lambda t} + 24e^{-17\lambda t}$$
$$- 4e^{-14\lambda t} - 20e^{-18\lambda t} + 32e^{-19\lambda t} - 24e^{-20\lambda t} + 8e^{-21\lambda t} - e^{-22\lambda t} \tag{4.108}$$

$$R_{t-N\times \text{NPBN}}(t) = 1 - \left(1 - e^{-\lambda t}\left(R_{t-\frac{N}{2}\times\frac{N}{2}\text{Benes}}(t)\right)\right)^2 \tag{4.109}$$

$$R_{t-8\times 8\ \text{RBN}}(t) = 8e^{-5\lambda t} - 4e^{-6\lambda t} - 8e^{-8\lambda t} + 8e^{-9\lambda t} - 18e^{-10\lambda t} + 16e^{-11\lambda t}$$
$$+ 32e^{-13\lambda t} - 48e^{-14\lambda t} + 24e^{-15\lambda t} - 4e^{-12\lambda t}$$
$$- 20e^{-16\lambda t} + 32e^{-17\lambda t} - 24e^{-18\lambda t} + 8e^{-19\lambda t} - e^{-20\lambda t} \tag{4.110}$$

$$R_{t-N\times \text{NRBN}}(t) = 1 - (1 - (R_{t-N\times N\ \text{Benes}}(t)))^2 \tag{4.111}$$

Based on the equations mentioned above, as well as considering the hypothesis that a logical calculation for λ is about 10^{-6} per hour [9, 20, 52], the time-dependent terminal reliability examination as a function of time for the network sizes of 8, 16, and 32 is shown in Figs. 4.64, 4.65, and 4.66, in order. Very similar outcomes can be obtained from these diagrams. Based on the figures, as is predicted, the terminal reliability decreases as the time increases. Nonetheless, a better execution of a network will be attained given less reduction in comparison to the other networks. Considering the charts, the weakest outcomes belong to two networks; EBN and Benes. The terminal reliability of these two networks is very similar to each other for different times. Nevertheless, Benes achieves a rather higher terminal reliability compared to EBN. As a result, the approach of having more stages is not efficient in order to enhance the reliability of the Benes network.

Based on the outcomes, the two networks, PBN and RBN, achieved the highest terminal reliability compared to the other networks. However, RBN is more considerable as higher terminal reliability can be achieved with RBN compared to PBN through various times. However, as the figures demonstrate, this is not a desirable outcome in comparison to PBN.

D. *Broadcast reliability (time-dependent analysis)*

Time-dependent broadcast reliability is defined as the probability of successful connection from a given source node to all destination nodes as a function of time.

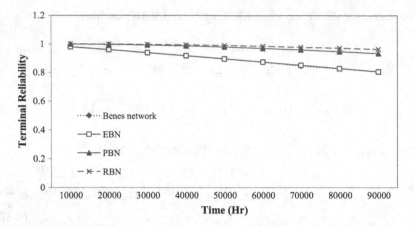

Fig. 4.64 Terminal reliability as a function of time in network size 8

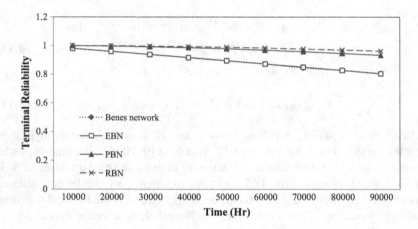

Fig. 4.65 Terminal reliability as a function of time in network size 16

It is obvious that time has a negative influence on reliability trend. However, it certainly would be practical to examine the influence. Here, like the previous subsection, it is assumed that the time to failure of the switching components is described with an exponential distribution and a good estimate for λ is about 10^{-6} per hour.

The broadcast reliability of Benes network, EBN, PBN, and RBN can be calculated by following equations:

$$R_{b-8\times 8 \text{ Benes}}(t) = 4e^{-9\lambda t} - 2e^{-10\lambda t} - 4e^{-13\lambda t} + 4e^{-14\lambda t} - e^{-15\lambda t} \qquad (4.112)$$

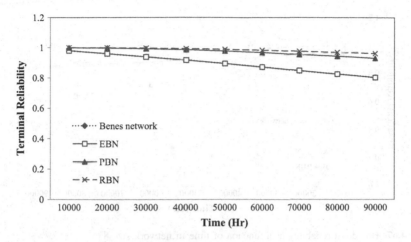

Fig. 4.66 Terminal reliability as a function of time in network size 32

$$R_{b-N \times N \text{ Benes}}(t) = e^{-\frac{N+2}{2}\lambda t} \left[1 - \left(1 - \left(R_{b-\frac{N}{2} \times \frac{N}{2} \text{Benes}}(t) \right) \right)^2 \right] \quad (4.113)$$

$$R_{b-8 \times 8 \text{ EBN}}(t) = 8e^{-10\lambda t} - 8e^{-11\lambda t} + 2e^{-12\lambda t} - 16c^{-15\lambda t} + 32e^{-16\lambda t}$$
$$- 24e^{-17\lambda t} + 8e^{-18\lambda t} - e^{-19\lambda t} \quad (4.114)$$

$$R_{b-N \times N \text{ EBN}}(t) = e^{-\frac{N+2}{2}\lambda t} \left[1 - \left(1 - \left(R_{b-\frac{N}{2} \times \frac{N}{2} \text{ EBN}}(t) \right) \right)^2 \right] \quad (4.115)$$

$$R_{b-16 \times 16 \text{ PBN}}(t) = 8e^{-\frac{27}{2}\lambda t} - 4e^{-\frac{29}{2}\lambda t} - 8e^{-\frac{35}{2}\lambda t} + 8e^{-\frac{37}{2}\lambda t} - 2e^{-\frac{39}{2}\lambda t} - 16e^{-27\lambda t}$$
$$+ 16e^{-28\lambda t} - 4e^{-29\lambda t} + 32e^{-31\lambda t} - 48e^{-32\lambda t} + 24e^{-33\lambda t} - 4e^{-34\lambda t}$$
$$- 16e^{-35\lambda t} + 32e^{-36\lambda t} - 24e^{-37\lambda t} + 8e^{-38\lambda t} - e^{-39\lambda t}$$
$$(4.116)$$

$$R_{b-N \times N \text{ PBN}}(t) = 1 - \left(1 - \left(e^{-\frac{N+2}{4}\lambda t} \left(R_{b-\frac{N}{2} \times \frac{N}{2} \text{Benes}}(t) \right) \right) \right)^2 \quad (4.117)$$

$$R_{b-8 \times 8 \text{ RBN}}(t) = 8e^{-9\lambda t} - 4e^{-10\lambda t} - 8e^{-13\lambda t} + 8e^{-14\lambda t} - 2e^{-15\lambda t} - 16e^{-18\lambda t}$$
$$+ 16e^{-19\lambda t} - 4e^{-20\lambda t} + 32e^{-22\lambda t} - 48e^{-23\lambda t} + 24e^{-24\lambda t} - 4e^{-25\lambda t}$$
$$- 16e^{-26\lambda t} + 32e^{-27\lambda t} - 24e^{-28\lambda t} + 8e^{-29\lambda t} - e^{-30\lambda t}$$
$$(4.118)$$

$$R_{b-N \times N \text{ RBN}}(t) = 1 - \left(1 - \left(R_{b-N \times N \text{ Benes}}(t) \right) \right)^2 \quad (4.119)$$

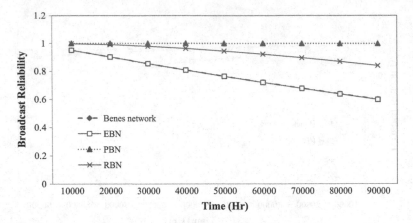

Fig. 4.67 Broadcast reliability as a function of time in network size 8

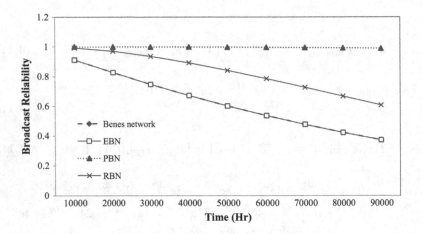

Fig. 4.68 Broadcast reliability as a function of time in network size 16

Figures 4.67, 4.68, and 4.69 show time-dependent broadcast reliability examination outcomes for network sizes of 8, 16, and 32, respectively.

Based on the outcomes, it is obvious that the highest broadcast reliability in various times is achieved by PBN, compared to other networks. By considering each figure independently, it can be concluded that as time increases, it becomes more beneficial to decide PBN as the dominant network. Also, considering all the outcomes of these figures alongside each other, it can be argued that as network size increases, PBN is the most beneficial of all networks again. The obtained outcomes reflect the idea that utilizing multiple parallel Benes networks is much more beneficial than the other approaches regarding broadcast reliability.

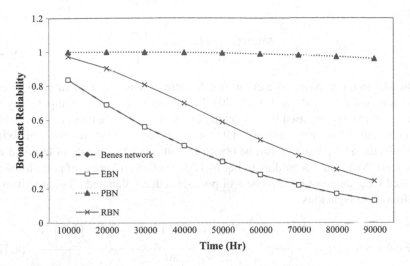

Fig. 4.69 Broadcast reliability as a function of time in network size 32

Opposed to the former subsection in which RBN leads to a less beneficial advantage in comparison to the PBN regarding terminal reliability, in this analysis, PBN achieves a clear and extraordinary broadcast reliability in comparison to RBN.

Totally, based on the outcomes achieved in this examination and the previous one, PBN achieves rather weaker outcomes regarding terminal reliability than RBN. Nevertheless, RBN's logical superiority regarding its broadcast reliability makes us to take the PBN as the best enhanced Benes network. In addition, considering all the discussions in this section, it can be argued that the method of utilizing multiple Benes networks in parallel is a better option to improve reliability compared to the other two approaches of increasing more stages and using replicated Benes networks.

E. Cost-effectiveness

Another concern of researchers during design of high-reliable interconnection networks is the cost of hardware. In other words, redundancy in hardware devices can enhance the reliability of the network. However, the creation of redundancy resulted in an increase in cost that must be acceptable and proportionate in terms of enhancement in reliability. Thus, the cost-effectiveness metric would be the metric that can be utilized in analysis of the MINs performance referring to cost. Also, this metric has been emphasized in the most pervious works [1, 6, 9, 52, 53]. Therefore, this crucial metric will be analyzed as further examination will be carried out regarding the performance of the networks. The cost-effectiveness (CE) parameter is determined by Eq. (4.120) [1, 6, 9, 52, 53].

$$CE = \frac{\text{Mean time to failure}}{\text{Cost}} = \frac{\int_0^\infty R(t)\,dt}{\text{Cost}} \qquad (4.120)$$

In addition, one common method is calculating crosspoint cost in order to estimate the cost of a MIN in Eq. (4.120). The crosspoint cost is computed by the number of crosspoints used in a switching element and by the number of switching elements used in the network [10, 49]. Also, similar to the reliability, cost-effectiveness parameter can be obtained from two standpoints of terminal and broadcast. As a result, according to Eq. (4.120), cost-effectiveness of point-terminal, denoted CE_T, and cost-effectiveness of point-broadcast, denoted CE_B, are given by the following equations:

$$CE_T(8 \times 8\,\text{Benes}) = \frac{\int_0^\infty \left(e^{-2\lambda t}\left[1 - \left(1 - \left(e^{-2\lambda t}\left(1 - (1 - e^{-\lambda t})^2\right)\right)\right)^2\right]\right)dt}{80} \qquad (4.121)$$

$$CE_T(N \times N\,\text{Benes}) = \frac{\int_0^\infty \left(e^{-2\lambda t}\left[1 - \left(1 - \left(R_{t-\frac{N}{2}\times\frac{N}{2}\text{Benes}}(t)\right)\right)^2\right]\right)dt}{2N(2(\log_2 N) - 1)} \qquad (4.122)$$

$$CE_T(8 \times 8\,\text{EBN}) = \frac{\int_0^\infty \left(e^{-2\lambda t}\left[1 - \left(1 - \left(e^{-2\lambda t}\left(1 - (1 - e^{-\lambda t})^2\right)^2\right)\right)^2\right]\right)dt}{96} \qquad (4.123)$$

$$CE_T(N \times N\,\text{EBN}) = \frac{\int_0^\infty \left(e^{-2\lambda t}\left[1 - \left(1 - \left(R_{t-\frac{N}{2}\times\frac{N}{2}\text{EBN}}(t)\right)\right)^2\right]\right)dt}{2N(2(\log_2 N))} \qquad (4.124)$$

$$CE_T(16 \times 16\,\text{PBN}) = \frac{\int_0^\infty \left(1 - \left(1 - \left(e^{-3\lambda t}\left[1 - \left(1 - \left(e^{-2\lambda t}\left(1 - (1 - e^{-\lambda t})^2\right)\right)\right)^2\right]\right)\right)^2\right)dt}{224} \qquad (4.125)$$

$$CE_T(N \times N\,\text{PBN}) = \frac{\int_0^\infty \left(1 - \left(1 - e^{-\lambda t}\left(R_{t-\frac{N}{2}\times\frac{N}{2}\text{Benes}}(t)\right)\right)^2\right)dt}{2N(2 + ((2\log_2 N) - 3))} \qquad (4.126)$$

$$CE_T(8 \times 8\,\text{RBN}) = \frac{\int_0^\infty \left(1 - \left(1 - \left(e^{-2\lambda t}\left[1 - \left(1 - \left(e^{-2\lambda t}\left(1 - (1 - e^{-\lambda t})^2\right)\right)\right)^2\right]\right)\right)^2\right)dt}{160} \qquad (4.127)$$

$$\text{CE}_\text{T}(N \times N \text{ RBN}) = \frac{\int\limits_{0}^{\infty} \left(1 - (1 - (R_{t-N \times N \text{ Benes}}(t)))^2\right) dt}{4N(2(\log_2 N) - 1)} \tag{4.128}$$

$$\text{CE}_\text{B}(8 \times 8 \text{ Benes}) = \frac{\int\limits_{0}^{\infty} \left(e^{-5\lambda t}\left[1 - \left(1 - \left(e^{-3\lambda t}\left(1 - (1 - e^{-\lambda t})^2\right)\right)\right)^2\right]\right) dt}{80} \tag{4.129}$$

$$\text{CE}_\text{B}(N \times N \text{ Benes}) = \frac{\int\limits_{0}^{\infty} \left(e^{-\frac{N+2}{2}\lambda t}\left[1 - \left(1 - \left(R_{b-\frac{N}{2} \times \frac{N}{2}\text{Benes}}(t)\right)^2\right)\right]\right) dt}{2N(2(\log_2 N) - 1)} \tag{4.130}$$

$$\text{CE}_\text{B}(8 \times 8 \text{ EBN}) = \frac{\int\limits_{0}^{\infty} \left(e^{-5\lambda t}\left[1 - \left(1 - \left(e^{-3\lambda t}\left(1 - (1 - e^{-\lambda t})^2\right)^2\right)\right)^2\right]\right) dt}{96} \tag{4.131}$$

$$\text{CE}_\text{B}(N \times N \text{ EBN}) = \frac{\int\limits_{0}^{\infty} \left(e^{-\frac{N+2}{2}\lambda t}\left[\left(1 - \left(R_{b-\frac{N}{2} \times \frac{N}{2}\text{EBN}}(t)\right)\right)^2\right]\right) dt}{2N(2(\log_2 N))} \tag{4.132}$$

$$\text{CE}_\text{B}(16 \times 16 \text{ PBN}) = \frac{\int\limits_{0}^{\infty} \left(1 - \left(1 - \left(e^{-\frac{19}{2}\lambda t}\left[1 - \left(1 - \left(e^{-3\lambda t}\left(1 - (1 - e^{-\lambda t})^2\right)\right)\right)^2\right]\right)\right)^2\right) dt}{224} \tag{4.133}$$

$$\text{CE}_\text{B}(N \times N \text{ PBN}) = \frac{\int\limits_{0}^{\infty} \left(1 - \left(1 - \left(e^{-\frac{N+2}{4}\lambda t}\left(R_{b-\frac{N}{2} \times \frac{N}{2}\text{Benes}}(t)\right)\right)\right)^2\right) dt}{2N(2 + ((2\log_2 N) - 3))} \tag{4.134}$$

$$\text{CE}_\text{B}(8 \times 8 \text{ RBN}) = \frac{\int\limits_{0}^{\infty} \left(1 - \left(1 - \left(e^{-5\lambda t}\left[1 - \left(1 - \left(e^{-3\lambda t}\left(1 - (1 - e^{-\lambda t})^2\right)\right)\right)^2\right]\right)\right)^2\right) dt}{160} \tag{4.135}$$

$$\text{CE}_\text{B}(N \times N \text{ RBN}) = \frac{\int\limits_{0}^{\infty} \left(1 - (1 - (R_{b-N \times N \text{ Benes}}(t)))^2\right) dt}{4N(2(\log_2 N) - 1)} \tag{4.136}$$

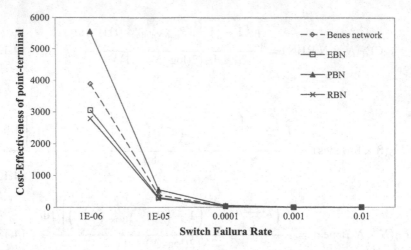

Fig. 4.70 Cost-effectiveness of point-terminal as a function of switch failure rate in network size 8

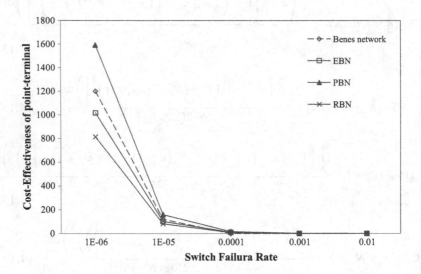

Fig. 4.71 Cost-effectiveness of point-terminal as a function of switch failure rate in network size 16

Based on Eqs. (4.121–4.128), Figs. 4.70, 4.71, and 4.72 show the outcomes of cost-effectiveness of point-terminal thorough examination as a function of the switch failure rate (λ) for network sizes 8, 16, and 32, in order. These figures demonstrate almost the similar outcomes. These outcomes show that the PBN is the most proper option regarding cost-effectiveness of point-terminal than other networks. In fact, the PBN can attain higher efficiency versus its increased cost.

In addition, based on Eqs. (4.129–4.136), the results of cost-effectiveness of point-broadcast thorough examination as a function of the switch failure rate (λ) for

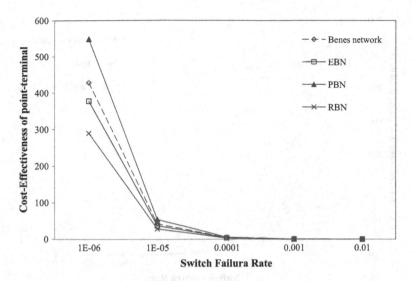

Fig. 4.72 Cost-effectiveness of point-terminal as a function of switch failure rate in network size 32

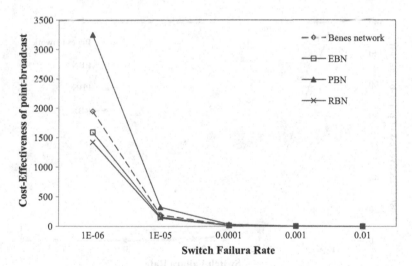

Fig. 4.73 Cost-effectiveness of point-broadcast as a function of switch failure rate in network size 8

network sizes 8, 16, and 32 are illustrated in Figs. 4.73, 4.74, and 4.75, respectively.

According to Figs. 4.73, 4.74, and 4.75, it is obvious that the most ideal and the worst outcomes regarding cost-effectiveness of point-broadcast are achieved by PBN and RBN, in order. In fact, based on the outcomes attained in the former and

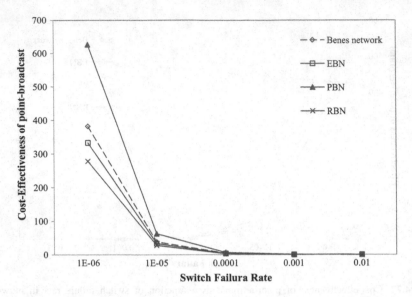

Fig. 4.74 Cost-effectiveness of point-broadcast as a function of switch failure rate in network size 16

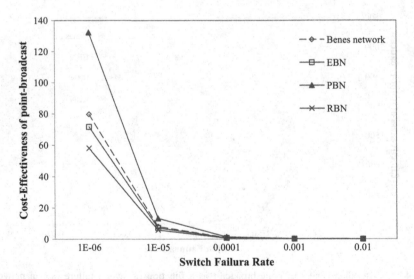

Fig. 4.75 Cost-effectiveness of point-broadcast as a function of switch failure rate in network size 32

this subsection, it has been argued that the method of using multiple Benes networks in parallel can meet more advantage to the Benes network regarding reliability and cost-effectiveness metrics compared to the approach of increasing stages to Benes network and utilizing replicated Benes networks.

Totally, time-independent evaluation demonstrated that both methodologies attain almost the same outcomes. Nevertheless, the approach of using replicated Benes networks is rather more considerable than utilizing multiple Benes networks in parallel regarding terminal reliability. In addition, the outcomes show that the idea of utilizing several Benes networks in parallel provides rather better execution than utilizing replicated Benes networks regarding broadcast reliability. As a result, achieving the most suitable remedy from both approaches was not easy in order to enhance the reliability of the Benes network. Therefore, the time-dependent analyses were helpful in selecting the most practical option. The time-dependent examination demonstrated even though the approach of utilizing multiple Benes networks in parallel was rather weaker than utilizing replicated Benes networks regarding terminal reliability, it attained a considerable broadcast reliability in comparison to the approach of using replicated Benes networks. Because of this and according to the vital role of broadcast communication in MINs, it is concluded that the most suitable approach in order to enhance the reliability of Benes network from the three important achievements to boost the reliability of the MINs is utilizing several Benes networks in parallel. In addition, the outcomes of the cost-effectiveness examination showed that utilizing several Benes networks in parallel is a cost-effective idea in comparison with other two approaches.

References

1. Blake JT, Trivedi KS (1989) Multistage interconnection network reliability. IEEE Trans Comput 38(11):1600–1604
2. Fan CC, Bruck J (2000) Tolerating multiple faults in multistage interconnection networks with minimal extra stages. IEEE Trans Comput 49(9):998–1004
3. Adams GB, Siegel HJ (1982) The extra stage cube: a fault-tolerant interconnection network for supersystems. IEEE Trans Comput 100(5):443–454
4. Koren I, Krishna CM (2007) Fault-tolerant systems. Morgan Kaufmann, USA
5. Veglis A, Pomportsis A (2001) Dependability evaluation of interconnection networks. Comput Electr Eng 27(3):239–263
6. Blake JT, Trivedi KS (1989) Reliability analysis of interconnection networks using hierarchical composition. IEEE Trans Reliab 38(1):111–120
7. Gunawan I (2008) Reliability analysis of shuffle-exchange network systems. Reliab Eng Syst Safety 93(2):271–276
8. Newman P (1989) Fast packet switching for integrated services. University of Cambridge, Computer Laboratory
9. Bansal PK, Joshi RC, Singh K (1994) On a fault-tolerant multistage interconnection network. Comput Electr Eng 20(4):335–345
10. Bistouni F, Jahanshahi M (2014) Improved extra group network: a new fault-tolerant multistage interconnection network. J Supercomputing 69(1):161–199
11. Cheema KK, Aggarwal R (2009) Design scheme and performance evaluation of a new fault-tolerant multistage interconnection network. Int J Comput Sci Netw Secur 9(9):270–276
12. Wu CL, Feng TY (1980) On a class of multistage interconnection networks. IEEE Trans Comput 100(8):694–702
13. Patel JH (1981) Performance of processor-memory interconnections for multiprocessors. IEEE Trans Comput 100(10):771–780

14. Siegel HJ, Smith SD (1978) Study of multistage SIMD interconnection networks. In: Proceedings of the 5th annual symposium on computer architecture. ACM
15. Kumar VP, Reddy SM (1985) Design and analysis of fault-tolerant multistage interconnection networks with low link complexity. In: ACM SIGARCH computer architecture news, vol. 13, No 3. IEEE Computer Society Press
16. Kumar VP, Reddy SM (1987) Augmented shuffle-exchange multistage interconnection networks. Computer 20(6):30–40
17. Wei S, Lee G (1988) Extra group network: a cost-effective fault-tolerant multistage interconnection network. In: ACM SIGARCH computer architecture news, vol 16, No 2. IEEE Computer Society Press
18. Bistouni F, Jahanshahi M (2014) Analyzing the reliability of shuffle-exchange networks using reliability block diagrams. Reliab Eng Syst Saf 132:97–106
19. Bansal PK, Singh K, Joshi RC (1993) Reliability and performance analysis of a modular multistage interconnection network. Microelectron Reliab 33(4):529–534
20. Aggarwal R, Kaur L (2008) On reliability analysis of fault-tolerant multistage interconnection networks. Int J Comput Sci Secur (IJCSS) 2(4):01–08
21. Gunawan I (2008) Redundant paths and reliability bounds in gamma networks. Appl Math Model 32(4):588–594
22. Fard NS, Gunawan I (2002) Reliability bounds for large multistage interconnection networks. In: Applied parallel computing. Springer, Berlin Heidelberg
23. Bhardwaj VP, Nitin N (2013) Message broadcasting via a new fault tolerant irregular advance omega network in faulty and nonfaulty network environments. J Electr Comput Eng 6
24. Sadawarti H, Bansal PK (2007) Fault tolerant irregular augmented shuffle network. In: Proceedings of the 2007 annual conference on international conference on computer engineering and applications. World Scientific and Engineering Academy and Society (WSEAS)
25. Das N, Mukhopadhyaya K, Dattagupta J (2000) O(n) routing in rearrangeable networks. J Syst Architect 46:529–542
26. Sheu TL, Lin W, Das CR (1995) Distributed fault diagnosis in multistage network-based multiprocessors. IEEE Trans Comput 44(9):1085–1095
27. Leung YW (1993) On-line fault identification in multistage interconnection networks. Parallel Comput 19(6):693–702
28. Chaki N, Bhattacharya S (2000) High level net models: a tool for permutation mapping and fault detection in multistage interconnection network. In: TENCON 2000. Proceedings, vol 2. IEEE
29. Choi M, Park N, Lombardi F (2003) Modeling and analysis of fault tolerant multistage interconnection networks. IEEE Trans Instrum Meas 52(5):1509–1519
30. Liu J et al (2014) Online traffic-aware fault detection for networks-on-chip. J Parallel Distrib Comput 74(1):1984–1993
31. Chakrabarty A, Collier M (2014) Routing algorithm for (2logN − 1)-stage switching networks and beyond. J Parallel Distrib Comput 74(10): 3045–3055
32. Kang WH, Kliese A (2014) A rapid reliability estimation method for directed acyclic lifeline networks with statistically dependent components. Reliab Eng Syst Saf 124:81–91
33. Kim Y, Kang WH (2013) Network reliability analysis of complex systems using a non-simulation-based method. Reliab Eng Syst Saf 110:80–88
34. Shuang Q, Zhang M, Yuan Y (2014) Node vulnerability of water distribution networks under cascading failures. Reliab Eng Syst Saf 124:132–141
35. Padmavathy N, Chaturvedi SK (2013) Evaluation of mobile ad hoc network reliability using propagation-based link reliability model. Reliab Eng Syst Saf 115:1–9
36. Jahanshahi M, Dehghan M, Meybodi MR (2013) LAMR: learning automata based multicast routing protocol for multi-channel multi-radio wireless mesh networks. Appl Intell 38(1):58–77

37. Jahanshahi M, Dehghan M, Meybodi MR (2013) On channel assignment and multicast routing in multi-channel multi-radio wireless mesh networks. Int J Ad Hoc Ubiquitous Comput 12(4):225–244
38. Jahanshahi M, Dehghan M, Meybodi MR (2011) A mathematical formulation for joint channel assignment and multicast routing in multi-channel multi-radio wireless mesh networks. J Netw Comput Appl 34(6):1869–1882
39. Jahanshahi M, Barmi AT (2014) Multicast routing protocols in wireless mesh networks: a survey. Computing 1–29
40. Jahanshahi M, Maddah M, Najafizadegan N (2013) Energy aware distributed partitioning detection and connectivity restoration algorithm in wireless sensor networks. Int J Math Model Comput 3(1):71–82
41. Jahanshahi M, Rahmani S, Ghaderi S (2013) An efficient cluster head selection algorithm for wireless sensor networks using fuzzy inference systems. Int J Smart Electr Eng (IJSEE) 2 (2):121–125
42. Ebrahimi N, McCullough K, Xiao Z (2013) Reliability of sensors based on nanowire networks operating in a dynamic environment. IEEE Trans Reliab 62(4):908–916
43. Schneider K et al (2013) Social network analysis via multi-state reliability and conditional influence models. Reliab Eng Syst Saf 109:99–109
44. Lin YK, Chang PC (2013) A novel reliability evaluation technique for stochastic-flow manufacturing networks with multiple production lines. IEEE Trans Reliab 62(1):92–104
45. Birolini A (2014) Reliability engineering: theory and practice. Springer, Berlin Heidelberg
46. Mettas A, Savva M (2001) System reliability analysis: the advantages of using analytical methods to analyze non-repairable systems. In: Reliability and maintainability symposium, 2001. Proceedings annual. IEEE
47. Aljundi AC et al (2006) A universal performance factor for multi-criteria evaluation of multistage interconnection networks. Future Gener Comput Syst 22(7):794–804
48. Cuda D, Giaccone P, Montalto M (2012) Design and control of next generation distribution frames. Comput Netw 56(13):3110–3122
49. Tutsch D, Hommel G (2008) MLMIN: a multicore processor and parallel computer network topology for multicast. Comput Oper Res 35(12):3807–3821
50. Yang Y, Wang J (2005) A new design for wide-sense nonblocking multicast switching networks. IEEE Trans Commun 53(3):497–504
51. Escudero-Sahuquillo J et al (2014) A new proposal to deal with congestion in InfiniBand-based fat-trees. J Parallel Distrib Comput 74(1):1802–1819
52. Bistouni F, Jahanshahi M (2015) Pars network: a multistage interconnection network with fault-tolerance capability. J Parallel Distrib Comput 75:168–183
53. Bistouni F, Jahanshahi M (2015) Scalable crossbar network: a non-blocking interconnection network for large-scale systems. J Supercomputing 71(2):697–728

Chapter 5
Scalable Crossbar Network

5.1 Introduction

In the previous chapter, we discussed fault-tolerant multistage interconnection networks. Fault-tolerant MINs can provide some significant performance criteria such as reliability, throughput, and cost-effectiveness. In addition, these networks are able to eliminate the problem of scalability in crossbar networks; besides, it can cause to diminish the blocking problem. However, these networks are not able to eliminate the blocking problem efficiently. Therefore, the remaining challenge for researchers in this field is designing networks that can completely solve the blocking problem. Such networks are so-called non-blocking networks.

Based on the previous discussion, in this chapter, we give an interconnection to the network that can solve the following issues: (1) Blocking problem which is very significant in this area (2) The scalability problem related to size of the largest crossbar implementable in a single chip.

A typical crossbar network (shown in Fig. 1.2) can be seen as a network containing only one switching element. As a matter of fact, only one switch is located between any pair of source–destination in this structure. This topology can lead to great benefits: Having just one switch between each pair of source–destination provides possibility of increasing the connection speed between the terminal nodes. Really, in this topology, path-length delay equals to one. In addition, this network is non-blocking for each permutation of the connections. The permutation here is intended for the request for simultaneous connections in all N inputs for their distinctive destination.

Another point is that the hardware cost can be computed by considering the number of crosspoints within each network. Consequently, the cost of a crossbar network equals to the number of crosspoints within a single switch, whose cost is less than MCN's hardware cost (shown in Fig. 3.12)—utilizing a large number of switches resulting in increased costs.

© Springer International Publishing AG, part of Springer Nature 2018   137
M. Jahanshahi and F. Bistouni, *Crossbar-Based Interconnection Networks*,
Computer Communications and Networks,
https://doi.org/10.1007/978-3-319-78473-1_5

Despite these positive features, the use of crossbar network in systems with large size is limited. This scalability problem arises from the restriction in the number of pins in a single-chip VLSI. The total number of pins in a chip should be in a particular level. However, by increasing the size of the crossbar network, more pins are required on a chip.

Based on the above discussions, the use of a single switch between all source–destination pairs is beneficial for the network, particularly in non-blocking state. Hence, we present a novel non-blocking interconnection network with this structure. This network is called scalable crossbar network (SCN). However, we need to solve the scalability problem, for which a number of small-size crossbars can be utilized instead of a single large-size crossbar. Also, every switch can be used to cover a particular group of sources and destinations. In this respect, we should use different network sources such as switches and links. Then, each group can be assigned to a specific set of terminal nodes. In fact, the layout of this new topology will be a combination of several methods to achieve the desired goals, like non-blocking, scalability and cost-effectiveness. These methods include: (1) using just one switch between every pair of source–destination similar to crossbar network (2) several small-scale crossbars usage as switches (3) Grouping of network resources to meet the following requirements: Fair allocation of resources to all terminal nodes on the network can facilitate the routing mechanism, maintaining just one switch between every source–destination pair.

The initial two ideas mentioned above are used in previous topologies: the first in the crossbar network and the second in scalable networks such as MINs and MCN. Nevertheless, as far as we know, the third idea is new and has never been used in any of the prior work for designing non-blocking networks. To mention another point, the combination of the above three methods is also a creativity in this work. In the following, we will learn more about the topological structure of the SCN.

5.2   Scalable Crossbar Network (SCN) Structure

A SCN of size 8×8 is shown in Fig. 5.1. As you can see, it is made of several 2×2 crossbar switches. In this figure, links to the source 0 are highlighted to better understand this structure. As a general case, consider a SCN of size $N \times N$. In this case, the number of switching groups equals to $\left(\frac{N}{C}\right)$ and each group is made up of $\left(\frac{N}{C}\right)$ crossbar switches of size $C \times C$. As an example, consider $(N = 8)$ and $(C = 2)$ as shown in Fig. 5.1. In this case, the SCN in size of 8×8 is required four switching sets. Each of them has four crossbar switches in size of 2×2. It should be considered that the groups are labeled as $G_i, i = 1, 2, \ldots, \left(\frac{N}{C}\right)$ and the switching elements are showed as $SE_{i,j}$, where i represents the group number such that $i = 1, 2, \ldots, \left(\frac{N}{C}\right)$ and j is the switch number such that $j = 1, 2, \ldots, \left(\frac{N}{C}\right)$.

In the SCN, each source node is linked to one crossbar switch in any group, conveying that any source is connected to $\left(\frac{N}{C}\right)$ crossbar switches in total. For

Fig. 5.1 A SCN of size
8 × 8

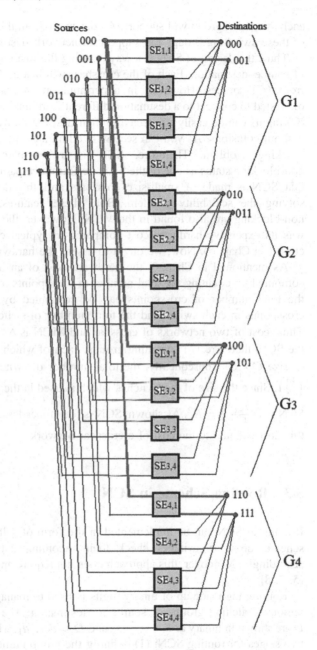

example, consider the source 000 (0) in Fig. 5.1. This source is connected to the
four switches $SE_{1,1}$, $SE_{2,1}$, $SE_{3,1}$, and $SE_{4,1}$. The point is that none of these switches
are connected to two different destinations compared to other switches. In this
example, $SE_{1,1}$ is connected to 000 and 001 destinations, while $SE_{2,1}$ to 010 and
011, and also $SE_{3,1}$ to 100 and 101, and finally $SE_{4,1}$ to 110 and 111. Generally,

each switching group will support (C) destinations in this topology. As a result, all of these switching groups can support all network destinations with each other.

Thus, the design of SCN is a way of using the just one switch between any pair of source–destination. Each of the crossbar switches in the network receives traffic from (C) sources. However, in a permutation of connections, each source is requested to connect to a destination different from any other source. As a result, all (C) inputs can be easily connected to their destination by $C \times C$ crossbar switches without collisions. As you can see, SCN eliminates the possible occurrence of the blocking problem. Therefore, SCN is a non-blocking network providing non-blocking status as one of the basic performance parameters. Another subject is that SCN is made of small-size crossbars and this designing technique allows solving the scalability problem. Both these features, namely scalability and non-blocking are also found in the MCN. However, the main problem with MCN was its expensive hardware cost compared to typical crossbar networks (as discussed in Chap. 3). Now, we ought to assess the hardware cost in the SCN.

As mentioned in Chap. 3, the hardware cost of an interconnection network is computed by counting the total number of crosspoints in the network. In addition, the total number of crosspoints can be computed by counting the number of crosspoints in each switch and the total number of switches in the network [1–6]. Thus, cost of two networks of crossbar and MCN is N^2 and $4N^2$, respectively. In the SCN, there are $\left(\frac{N}{C}\right)$ switching groups, each of which has $\left(\frac{N}{C}\right)$ crossbar switches of size $C \times C$. Consequently, the total number of switches in the SCN equals to $\left(\frac{N^2}{C^2}\right)$. Since the size of the switches which is used in the SCN is $C \times C$, the cost of SCN is $\left(\frac{N^2 \times C^2}{C^2}\right) = N^2$. As shown, SCN is four times less expensive than MCN, and this cost is equal to the cost of a crossbar network.

5.3   Routing Scheme in SCN

Routing in SCN can be implemented in the form of self-routing. In a self-routing scheme, crossbar switches check their incoming data and adjust themselves accordingly. Moreover, this approach does not require any central routing hardware [5, 7, 8].

Routing tag made up of binary digits is used to manage the connection through various switching groups of source to destination. The source S and destination D are shown in binary as $S = s_n \ldots s_1$ and $D = d_n \ldots d_1$, where $n = \log_2 N$. There are two stages for routing SCN: (1) defining the group number of switching (2) determining the state of a switch (straight or exchange).

In the SCN, the first step for sending messages from a source to destination is defining the switching group number (GN) for connected destination. By means of destination tag bits, GN is computed as follows:

$$GN = \left(\sum_{i=0}^{(\log_2 N)-2} d_{i+2}(2)^i \right) + 1 \tag{5.1}$$

As an instance, consider the destination 110 ($d_3 = 1, d_2 = 1, d_1 = 0$). The group number for this destination is calculated as

$$GN = \left(\sum_{i=0}^{1} d_{i+2}(2)^i \right) + 1 = \left(d_2(2)^0 + d_3(2)^1 \right) + 1 = (1+2) + 1 = 4 \tag{5.2}$$

After calculating the group number, determining the switch status is the next step. Switch status determines the output of the switch to which incoming message must be sent: upper or lower output. The status is called straight if the same output port receives incoming messages as the input switch port. Otherwise, the status of switch will be exchange. These two cases are shown in Fig. 4.2. In this figure, it is assumed that both input ports have incoming messages. The upper input and output lines are labeled as i and the lower input and output lines labeled as j. Then, straight and exchange statuses can be defined as: (1) straight: input i transferred to output i and input j transferred to output j; (2) exchange: reversely, input i transferred to output j and input j transferred to output i.

In SCN, switch statuses (SSs) are based on the source and destination tags bits, as follows:

$$SS = \begin{cases} \text{Straight,} & s_1 \oplus d_1 = 0 \\ \text{Exchange,} & s_1 \oplus d_1 = 1 \end{cases} \tag{5.3}$$

For example, consider the source 101 ($s_3 = 1, s_2 = 0, s_1 = 1$) and destination 100 ($d_3 = 1, d_2 = 0, d_1 = 0$), then we have

$$GN = \left(\sum_{i=0}^{1} d_{i+2}(2)^i \right) + 1 = \left(d_2(2)^0 + d_3(2)^1 \right) + 1 = (0+2) + 1 = 3 \tag{5.4}$$

Therefore, according to (5.4), the source connects to the third switched group. As $s_1 \oplus d_1 = 1 \oplus 0 = 1$, the switch status is exchanged.

For the better understanding, the values for source 000 of the route and all of its destinations are entered into Table 5.1.

For further discussion, the routing values to the permutation $P = \begin{pmatrix} 01234567 \\ 13572460 \end{pmatrix}$ will be:

Source 000 and destination 001:

$$GN = \left(\sum_{i=0}^{1} d_{i+2}(2)^i \right) + 1 = \left(d_2(2)^0 + d_3(2)^1 \right) + 1 = (0+0) + 1 = 1 \tag{5.5}$$

$$s_1 \oplus d_1 = 0 \oplus 1 = 1 \tag{5.6}$$

Table 5.1 Values for source 000 of the route

Source	Destinations	Group number	$(s_1 \oplus d_1)$	Switch state
000	000	1	0	Straight
	001	1	1	Exchange
	010	2	0	Straight
	011	2	1	Exchange
	100	3	0	Straight
	101	3	1	Exchange
	110	4	0	Straight
	111	4	1	Exchange

Source 001 and destination 011:

$$GN = \left(\sum_{i=0}^{1} d_{i+2}(2)^i \right) + 1 = \left(d_2(2)^0 + d_3(2)^1 \right) + 1 = (1+0) + 1 = 2 \quad (5.7)$$

$$s_1 \oplus d_1 = 1 \oplus 1 = 0 \quad (5.8)$$

Source 010 and destination 101:

$$GN = \left(\sum_{i=0}^{1} d_{i+2}(2)^i \right) + 1 = \left(d_2(2)^0 + d_3(2)^1 \right) + 1 = (0+2) + 1 = 3 \quad (5.9)$$

$$s_1 \oplus d_1 = 0 \oplus 1 = 1 \quad (5.10)$$

Source 011 and destination 111:

$$GN = \left(\sum_{i=0}^{1} d_{i+2}(2)^i \right) + 1 = \left(d_2(2)^0 + d_3(2)^1 \right) + 1 = (1+2) + 1 = 4 \quad (5.11)$$

$$s_1 \oplus d_1 = 1 \oplus 1 = 0 \quad (5.12)$$

Source 100 and destination 010:

$$GN = \left(\sum_{i=0}^{1} d_{i+2}(2)^i \right) + 1 = \left(d_2(2)^0 + d_3(2)^1 \right) + 1 = (1+0) + 1 = 2 \quad (5.13)$$

$$s_1 \oplus d_1 = 0 \oplus 0 = 0 \quad (5.14)$$

Source 101 and destination 100:

$$GN = \left(\sum_{i=0}^{1} d_{i+2}(2)^i \right) + 1 = \left(d_2(2)^0 + d_3(2)^1 \right) + 1 = (0+2) + 1 = 3 \quad (5.15)$$

$$s_1 \oplus d_1 = 1 \oplus 0 = 1 \quad (5.16)$$

Source 110 and destination 110:

$$GN = \left(\sum_{i=0}^{1} d_{i+2}(2)^i \right) + 1 = \left(d_2(2)^0 + d_3(2)^1 \right) + 1 = (1+2) + 1 = 4 \quad (5.17)$$

$$s_1 \oplus d_1 = 0 \oplus 0 = 0 \quad (5.18)$$

Source 111 and destination 000:

$$GN = \left(\sum_{i=0}^{1} d_{i+2}(2)^i \right) + 1 = \left(d_2(2)^0 + d_3(2)^1 \right) + 1 = (0+0) + 1 = 1 \quad (5.19)$$

$$s_1 \oplus d_1 = 1 \oplus 0 = 1 \quad (5.20)$$

5.4   The Analysis of Performance

In the next part, the performance for SCN will be analyzed comparing to recognized networks called SEN, SEN+, Benes network, two-layer replicated MIN [9], multilayer MIN 1248, multilayer MIN 1888 (the digits indicate the numbers of layers at stage 1, stage 2, etc). Moreover, the metrics for start of replication factor G_S, growth factor G_F, and limit factors of layer G_L would be described for each network independently. For example, in network 1248, $G_S = 2$, $G_F = 2$, and $G_L = 8$, and for the other network 1888, $G_S = 2$, $G_F = 8$, and $G_L = 8$.), and MCN [10]. We chose these networks to cover a wide range of various topologies. The SEN is a blocking network. Also, SEN+ network has fault-tolerant feature. The Benes is a network, which can be arranged again. The replicated MIN is a network with particular structure and high effectiveness. Moreover, according to [6], multilayer networks 1248 and 1888 provide the acceptable performance. In addition, the MCN is a non-blocking network.

In general, IEEE describes the feature of reliability as "The ability of a system or component to accomplish its essential tasks in specified conditions for a given period of time" [11]. Therefore, many researchers believe that the reliability metric is very important for each efficient network system in interconnection networks [1, 5, 12–22]. Since reliability analysis is a mathematical description of a system, it provides precise information about the system performance. An interconnection network must have the ability to guarantee reliable delivery of data. Reliable interconnection networks are designed in such a way that they continue to function even in case of occurrence of certain number of faults. In fact, the reliability of an interconnection network is a measure of its performance to deliver messages correctly. The delivery of messages without loss in 100% of the cases is a fundamental requirement.

Obviously, reliability is a key parameter in network performance. Moreover, for accurate determination of reliability in a system, the development of analytical methods for evaluating reliability is a must. Therefore, the reliability metric will be evaluated in detail in this section by analytical method. However, terminal reliability is a serious measure for reliability, which is a hot subject for many researchers [5, 12, 16, 17, 20]. Thus, in this division, our focus is estimating the terminal reliability.

The availability time of system is an important parameter. Also, in information technology, this parameter is often referred to "uptime." The duration of time for a system that is online among outages or failures for the system can be considered as the "time to failure." The mean time to failure (MTTF) is the average of the time to failure, or briefly speaking, the MTTF is the estimated value of the time to failure. We focus in this subject as the key performance parameter because of its definitive nature.

Mostly, the calculation of a failure distribution in the entire system using the failure rate of its components is the primary goal of performance analysis of that system. The parameter to be evaluated to consider these items is referred to system failure rate. Therefore, this influential parameter will be analyzed in this section as well for a further study of the performance of the networks.

Furthermore, the cost-effectiveness parameter has been emphasized in the most reported works [1, 6, 13, 14, 23] which can be used for the assessment of the MINs performance in terms of cost. So, this parameter will be considered in hardware cost estimation of the networks.

At this point, the model of switch fault is used for analyzing the reliability of the MINs. Also, it is assumed that any switching element (i.e., switching components, multiplexers, and demultiplexers) might be failed. Moreover, we suppose that the time to failure (TTF) of the switching components is defined with characteristics of the Weibull distribution. The most general expression of the Weibull is given by the three-parameter Weibull distribution expression includes life time variable t, scale parameter η, and Weibull slope or shape parameter β. Weibull distribution is one of the most widely utilized distributions in reliability.

5.4.1 Mathematical Analysis

In general, the terminal reliability, $R(t)$, is the possibility of successful connection from each given source to each specified destination.

In this section, we assume that the $r(t)$ is the probability of a 2×2 switching element ($SE_{2 \times 2}$) being operational. Furthermore, in switching components with different sizes, based on number of the gates, their operational probability can be computed with $r(t)$ [1, 6, 13]. Also, it is supposed that the complexity of hardware for a component is relative to the number of gates directly [1, 6].

In the SCN, there is only one switching element between each source–destination pair. Consequently,

$$R_{SCN}(t) = r(t) \tag{5.21}$$

Also, for the SEN and SEN+, we have:

$$R_{SEN}(t) = r(t)^{(\log_2 N)} \tag{5.22}$$

$$R_{SEN+}(t) = r(t)^2 \left(1 - \left(1 - r(t)^{((\log_2 N)-1)} \right)^2 \right) \tag{5.23}$$

Terminal reliability of the 8×8 Benes network is calculated as follows:

$$R_{8 \times 8Benes}(t) = r(t)^2 \left(1 - \left(1 - \left(r(t)^2 \left(1 - (1 - r(t))^2 \right) \right) \right)^2 \right) \tag{5.24}$$

Terminal reliability of the $N \times N$ Benes network is given by:

$$R_{N \times NBenes}(t) = r(t)^2 \left(1 - \left(\left(1 - \left(R_{\frac{N}{2} \times \frac{N}{2}Benes}(t) \right) \right)^2 \right) \right) \tag{5.25}$$

Moreover, we have the following formulas for the two-layer replicated MIN and multilayer MIN:

$$R_{replicated}(t) = r(t) \left(1 - \left(1 - r(t)^{(\log_2 N)} \right)^2 \right) \tag{5.26}$$

$$R_{network1248}(t) = r(t)^4 \left(1 - \left(1 - \left(r(t)^2 \left(1 - \left(1 - \left(r(t)^2 (1 - \left(1 - r(t)^{((\log_2 N)-3)} \right)^2 \right) \right)^2 \right) \right)^2 \right) \tag{5.27}$$

$$R_{network1888}(t) = r(t)^{10} \left(1 - \left(1 - r(t)^{((\log_2 N)-1)} \right)^8 \right) \tag{5.28}$$

As discussed before, (in Chap. 3), the path length for MCN (such as the number of switches from the source node to the destination node) among the various source–destination pairs is not the same one we have in this topology. Based on switching elements, that is between 1 and $(2N - 1)$, the path length may differ in this network. This relation leads to different reliabilities between different sources and destinations. To cope with this issue, we consider the average of path lengths between each source–destination pair. Actually, the path length of the source–destination pairs equivalents to $\frac{(2N-1)+1}{2} = N$. On the other hand, by this assuming, two dissimilar cases are probable in terminal reliability analysis. Firstly, we compute all of two cases; furthermore, we achieve the average of theme. Those two mentioned instances contain:

$$R_{\text{MCN}}(t) = r(t)^N \qquad (5.29)$$

$$R_{\text{MCN}}(t) = r(t)\left(r(t)^2 \left(1 - \left(1 - r(t)^{(N-3)} \right)^2 \right) \right) + (1 - r(t))r(t)^N \qquad (5.30)$$

In relation to (5.29) and (5.30), the average of terminal reliability is obtained by Eq. (5.31) for MCN.

$$R_{\text{MCN}}(t) = \frac{r(t)^N + r(t)\left(r(t)^2 \left(1 - \left(1 - r(t)^{(N-3)} \right)^2 \right) \right) + (1 - r(t))r(t)^N}{2} \qquad (5.31)$$

As previously mentioned, it is assumed that the times-to-failures of the switching components are described using Weibull life distribution. Therefore, we have:

$$R_{\text{SCN}}(t) = e^{-\left(\frac{t}{\eta}\right)^\beta} \qquad (5.32)$$

$$R_{\text{SEN}}(t) = e^{-(\log_2 N)\left(\frac{t}{\eta}\right)^\beta} \qquad (5.33)$$

$$R_{\text{SEN}+}(t) = e^{-2\left(\frac{t}{\eta}\right)^\beta}\left(1 - \left(1 - e^{-((\log_2 N)-1)\left(\frac{t}{\eta}\right)^\beta} \right)^2 \right) \qquad (5.34)$$

$$R_{8\times 8\text{Benes}}(t) = e^{-2\left(\frac{t}{\eta}\right)^\beta}\left(1 - \left(1 - \left(e^{-2\left(\frac{t}{\eta}\right)^\beta}\left(1 - \left(1 - e^{-\left(\frac{t}{\eta}\right)^\beta} \right)^2 \right) \right)^2 \right) \right)^2 \qquad (5.35)$$

$$R_{N\times N\text{Benes}}(t) = e^{-2\left(\frac{t}{\eta}\right)^\beta}\left(1 - \left(\left(1 - \left(R_{\frac{N}{2}\times\frac{N}{2}\text{Benes}}(t) \right) \right)^2 \right) \right) \qquad (5.36)$$

$$R_{\text{replicated}}(t) = e^{-\left(\frac{t}{\eta}\right)^\beta}\left(1 - \left(1 - e^{-(\log_2 N)\times\left(\frac{t}{\eta}\right)^\beta} \right)^2 \right) \qquad (5.37)$$

$$R_{\text{network1248}}(t) = e^{-4\left(\frac{t}{\eta}\right)^\beta}\left(1 - \left(1 - \left(e^{-2\left(\frac{t}{\eta}\right)^\beta}\left(1 - \left(1 - \left(e^{-2\left(\frac{t}{\eta}\right)^\beta}\left(1 - \left(1 - e^{-((\log_2 N)-3)\left(\frac{t}{\eta}\right)^\beta} \right)^2 \right) \right)^2 \right) \right)^2 \right) \right)^2 \right)$$

$$(5.38)$$

$$R_{\text{network1888}}(t) = e^{-10\left(\frac{t}{\eta}\right)^{\beta}}\left(1 - \left(1 - e^{-((\log_2 N)-1)\left(\frac{t}{\eta}\right)^{\beta}}\right)^{8}\right)$$

(5.39)

$$R_{\text{MCN}}(t) = \frac{e^{-N\left(\frac{t}{\eta}\right)^{\beta}} + e^{-\left(\frac{t}{\eta}\right)^{\beta}}\left(e^{(-2)\left(\frac{t}{\eta}\right)^{\beta}}\left(1 - \left(1 - e^{(-(N-3))\left(\frac{t}{\eta}\right)^{\beta}}\right)^{2}\right)\right) + \left(1 - e^{-\left(\frac{t}{\eta}\right)^{\beta}}\right)e^{(-N)\left(\frac{t}{\eta}\right)^{\beta}}}{2}$$

(5.40)

The MTTF is computed via the below equation:

$$\text{MTTF} = \int_{0}^{\infty} R(t)\,\mathrm{d}t$$

(5.41)

Accordingly, we can obtain the MTTF of the networks with the following equations:

$$\text{MTTF}_{\text{SCN}} = \int_{0}^{\infty}\left(e^{-\left(\frac{t}{\eta}\right)^{\beta}}\right)\mathrm{d}t$$

(5.42)

$$\text{MTTF}_{\text{SEN}} = \int_{0}^{\infty}\left(e^{-(\log_2 N)\left(\frac{t}{\eta}\right)^{\beta}}\right)\mathrm{d}t$$

(5.43)

$$\text{MTTF}_{\text{SEN}+} = \int_{0}^{\infty}\left(e^{-2\left(\frac{t}{\eta}\right)^{\beta}}\left(1 - \left(1 - e^{-((\log_2 N)-1)\left(\frac{t}{\eta}\right)^{\beta}}\right)^{2}\right)\right)\mathrm{d}t$$

(5.44)

$$\text{MTTF}_{8\times 8\text{Benes}} = \int_{0}^{\infty}\left(e^{-2\left(\frac{t}{\eta}\right)^{\beta}}\left(1 - \left(1 - \left(e^{-2\left(\frac{t}{\eta}\right)^{\beta}}\left(1 - \left(1 - e^{-\left(\frac{t}{\eta}\right)^{\beta}}\right)^{2}\right)\right)\right)^{2}\right)\right)\mathrm{d}t$$

(5.45)

$$\text{MTTF}_{N\times N\text{Benes}} = \int_{0}^{\infty}\left(e^{-2\left(\frac{t}{\eta}\right)^{\beta}}\left(1 - \left(\left(1 - \left(R_{\frac{N}{2}\times\frac{N}{2}\text{Benes}}(t)\right)\right)^{2}\right)\right)\right)\mathrm{d}t$$

(5.46)

$$\text{MTTF}_{\text{replicated}} = \int_{0}^{\infty}\left(e^{-\left(\frac{t}{\eta}\right)^{\beta}}\left(1 - \left(1 - e^{-(\log_2 N)\left(\frac{t}{\eta}\right)^{\beta}}\right)^{2}\right)\right)\mathrm{d}t$$

(5.47)

$$\text{MTTF}_{\text{MCN}} = \int_{0}^{\infty}\left(\frac{e^{-\left(N\left(\frac{t}{\eta}\right)\right)^{\beta}} + e^{-\left(\frac{t}{\eta}\right)^{\beta}}\left(e^{(-2)\left(\frac{t}{\eta}\right)^{\beta}}\left(1 - \left(1 - e^{(-(N-3))\left(\frac{t}{\eta}\right)^{\beta}}\right)^{2}\right)\right) + \left(1 - e^{-\left(\frac{t}{\eta}\right)^{\beta}}\right)e^{(-N)\left(\frac{t}{\eta}\right)^{\beta}}}{2}\right)\mathrm{d}t$$

(5.48)

$$
\mathrm{MTTF_{network1248}} = \int\limits_0^\infty \left(e^{-4\left(\frac{t}{\eta}\right)^\beta} \left(1 - \left(1 - \left(e^{-2\left(\frac{t}{\eta}\right)^\beta} \left(1 - \left(1 - \left(e^{-2\left(\frac{t}{\eta}\right)^\beta} \left(1 \right. \right. \right. \right. \right. \right. \right.
$$

$$
\left. \left. \left. \left. \left. \left. \left. - \left(1 - e^{-((\log_2 N)-3)\left(\frac{t}{\eta}\right)^\beta} \right)^2 \right) \right)^2 \right) \right) \right)^2 \right) \right) dt
$$

$$(5.49)$$

$$
\mathrm{MTTF_{network1888}} = \int\limits_0^\infty \left(e^{-10\left(\frac{t}{\eta}\right)^\beta} \left(1 - \left(1 - e^{-((\log_2 N)-1)\left(\frac{t}{\eta}\right)^\beta} \right)^8 \right) \right) dt \qquad (5.50)
$$

In addition, the cost-effectiveness (CE) parameter is calculated by Eq. (5.51) [1, 6, 13, 14, 23].

$$
\mathrm{CE} = \frac{\text{Mean time to failure}}{\text{Cost}} = \frac{\int_0^\infty R(t)\mathrm{d}t}{\text{Cost}} \qquad (5.51)
$$

Considering Eq. (5.51), we have:

$$
\mathrm{CE_{SCN}} = \frac{\int_0^\infty \left(e^{-\left(\frac{t}{\eta}\right)^\beta} \right) \mathrm{d}t}{N^2} \qquad (5.52)
$$

$$
\mathrm{CE_{SEN}} = \frac{\int_0^\infty \left(e^{-(\log_2 N)\left(\frac{t}{\eta}\right)^\beta} \right) \mathrm{d}t}{2N(\log_2 N)} \qquad (5.53)
$$

$$
\mathrm{CE_{SEN+}} = \frac{\int_0^\infty \left(e^{-2\left(\frac{t}{\eta}\right)^\beta} \left(1 - \left(1 - e^{-((\log_2 N)-1)\left(\frac{t}{\eta}\right)^\beta} \right)^2 \right) \right) \mathrm{d}t}{2N(\log_2 N + 1)} \qquad (5.54)
$$

$$
\mathrm{CE}_{N\times N\mathrm{Benes}} = \frac{\int_0^\infty \left(e^{-2\left(\frac{t}{\eta}\right)^\beta} \left(1 - \left(\left(1 - \left(R_{\frac{N}{2}\times\frac{N}{2}\mathrm{Benes}}(t) \right) \right)^2 \right) \right) \right) \mathrm{d}t}{2N(2(\log_2 N) - 1)} \qquad (5.55)
$$

$$
\mathrm{CE_{replicated}} = \frac{\int_0^\infty \left(e^{-\left(\frac{t}{\eta}\right)^\beta} \left(1 - \left(1 - e^{-(\log_2 N)\left(\frac{t}{\eta}\right)^\beta} \right)^2 \right) \right) \mathrm{d}t}{4N(\log_2 N + 1)} \qquad (5.56)
$$

$$
\mathrm{CE_{MCN}} = \frac{\int_0^\infty \left(\dfrac{e^{-\left(N\left(\frac{t}{\eta}\right)\right)^\beta} + e^{-\left(\frac{t}{\eta}\right)^\beta} \left(e^{(-2)\left(\frac{t}{\eta}\right)^\beta} \left(1 - \left(1 - e^{(-(N-3))\left(\frac{t}{\eta}\right)^\beta} \right)^2 \right) \right) + \left(1 - e^{-\left(\frac{t}{\eta}\right)^\beta} \right) e^{(-N)\left(\frac{t}{\eta}\right)^\beta}}{2} \right) \mathrm{d}t}{4N^2}
$$

$$(5.57)$$

$$CE_{network1248} = \frac{\int_0^\infty \left(e^{-4\left(\frac{t}{\eta}\right)^\beta} \left(1 - \left(1 - \left(e^{-2\left(\frac{t}{\eta}\right)^\beta} \left(1 - \left(1 - \left(e^{-2\left(\frac{t}{\eta}\right)^\beta} \left(1 - \left(1 - e^{-((\log_2 N)-3)\left(\frac{t}{\eta}\right)^\beta} \right)^2 \right) \right) \right) \right)^2 \right) \right) \right)^2 \right) dt}{36N + (16N(\log_2 N - 3))}$$

(5.58)

$$CE_{network1888} = \frac{\int_0^\infty \left(e^{-10\left(\frac{t}{\eta}\right)^\beta} \left(1 - \left(1 - e^{-((\log_2 N)-1)\left(\frac{t}{\eta}\right)^\beta} \right)^8 \right) \right) dt}{24N + (16N(\log_2 N - 1))}$$

(5.59)

However, the system failure rate, defined by λ, is computed by below equation [2, 20, 24].

$$\lambda = -\frac{1}{R(t)} \frac{d(R(t))}{dt}$$

(5.60)

According to Eq. (5.61), we have:

$$\lambda_{SCN} = \frac{-\left(-\frac{\beta\left(\frac{t}{\eta}\right)^{\beta-1}}{\eta e^{\left(\frac{t}{\eta}\right)^\beta}} \right)}{e^{-\left(\frac{t}{\eta}\right)^\beta}} = \frac{\beta\left(\frac{t}{\eta}\right)^{\beta-1}}{\eta}$$

(5.61)

$$\lambda_{SEN} = \frac{-\left(-\frac{\beta\log_2(N)\left(\frac{t}{\eta}\right)^{\beta-1}}{\eta e^{\log_2(N)\left(\frac{t}{\eta}\right)^\beta}} \right)}{e^{\left(-(\log_2(N))\right)\left(\frac{t}{\eta}\right)^\beta}} = \frac{\beta\log_2(N)\left(\frac{t}{\eta}\right)^{\beta-1}}{\eta}$$

(5.62)

$$\lambda_{SEN+} = \frac{e^{2\left(\frac{t}{\eta}\right)^\beta} \left(\frac{2\beta\left(\frac{t}{\eta}\right)^{\beta-1}\left(-\left(-\frac{1}{e^{(\log_2(N)-1)\left(\frac{t}{\eta}\right)^\beta}} + 1 \right)^2 + 1 \right)}{\eta e^{2\left(\frac{t}{\eta}\right)^\beta}} + \frac{2\beta(\log_2(N)-1)\left(\frac{t}{\eta}\right)^{\beta-1} e^{(-\log_2(N))\left(\frac{t}{\eta}\right)^\beta - \left(\frac{t}{\eta}\right)^\beta}\left(-\frac{1}{e^{(\log_2(N)-1)\left(\frac{t}{\eta}\right)^\beta}} + 1 \right)}{\eta} \right)}{-\left(-\frac{1}{e^{(\log_2(N)-1)\left(\frac{t}{\eta}\right)^\beta}} + 1 \right)^2 + 1}$$

(5.63)

$$\lambda_{8\times8Benes} = \frac{-\left(\frac{10\beta\left(\frac{t}{\eta}\right)^{\beta-1}}{\eta e^{10\left(\frac{t}{\eta}\right)^\beta}} + \frac{12\beta\left(\frac{t}{\eta}\right)^{\beta-1}}{\eta e^{6\left(\frac{t}{\eta}\right)^\beta}} + \frac{32\beta\left(\frac{t}{\eta}\right)^{\beta-1}}{\eta e^{8\left(\frac{t}{\eta}\right)^\beta}} - \frac{20\beta\left(\frac{t}{\eta}\right)^{\beta-1}}{\eta e^{5\left(\frac{t}{\eta}\right)^\beta}} - \frac{36\beta\left(\frac{t}{\eta}\right)^{\beta-1}}{\eta e^{9\left(\frac{t}{\eta}\right)^\beta}} \right)}{e^{(-2)\left(\frac{t}{\eta}\right)^\beta} \left(1 - \left(1 - \left(e^{(-2)\left(\frac{t}{\eta}\right)^\beta} \left(1 - \left(1 - e^{-\left(\frac{t}{\eta}\right)^\beta} \right)^2 \right) \right)^2 \right) \right)^2}$$

(5.64)

$$= \frac{e^{2\left(\frac{t}{\eta}\right)^\beta} \left(\frac{20\beta\left(\frac{t}{\eta}\right)^{\beta-1}}{\eta e^{5\left(\frac{t}{\eta}\right)^\beta}} + \frac{36\beta\left(\frac{t}{\eta}\right)^{\beta-1}}{\eta e^{9\left(\frac{t}{\eta}\right)^\beta}} - \frac{10\beta\left(\frac{t}{\eta}\right)^{\beta-1}}{\eta e^{10\left(\frac{t}{\eta}\right)^\beta}} - \frac{12\beta\left(\frac{t}{\eta}\right)^{\beta-1}}{\eta e^{6\left(\frac{t}{\eta}\right)^\beta}} - \frac{32\beta\left(\frac{t}{\eta}\right)^{\beta-1}}{\eta e^{8\left(\frac{t}{\eta}\right)^\beta}} \right)}{-\left(\frac{1}{e^{4\left(\frac{t}{\eta}\right)^\beta}} - \frac{2}{e^{3\left(\frac{t}{\eta}\right)^\beta}} + 1 \right)^2 + 1}$$

$$
\lambda_{\text{replicated}} = \frac{-\left(-\dfrac{\beta\left(\frac{t}{\eta}\right)^{\beta-1}\left(-\left(-\frac{1}{e^{\log_2(N)\left(\frac{t}{\eta}\right)^{\beta}}}+1\right)^2+1\right)}{\eta e^{\left(\frac{t}{\eta}\right)^{\beta}}} - \dfrac{2\beta\log_2(N)\left(\frac{t}{\eta}\right)^{\beta-1}e^{\left(-\log_2(N)\right)\left(\frac{t}{\eta}\right)^{\beta}-\left(\frac{t}{\eta}\right)^{\beta}}\left(-\frac{1}{e^{\log_2(N)\left(\frac{t}{\eta}\right)^{\beta}}}+1\right)}{\eta}\right)}{e^{-\left(\frac{t}{\eta}\right)^{\beta}}\left(1-\left(1-e^{\left(-\left(\log_2(N)\right)\right)\left(\frac{t}{\eta}\right)^{\beta}}\right)^2\right)}
$$

$$
= \frac{e^{\left(\frac{t}{\eta}\right)^{\beta}}\left(\dfrac{\beta\left(\frac{t}{\eta}\right)^{\beta-1}\left(-\left(-\frac{1}{e^{\log_2(N)\left(\frac{t}{\eta}\right)^{\beta}}}+1\right)^2+1\right)}{\eta e^{\left(\frac{t}{\eta}\right)^{\beta}}} + \dfrac{2\beta\log_2(N)\left(\frac{t}{\eta}\right)^{\beta-1}e^{\left(-\log_2(N)\right)\left(\frac{t}{\eta}\right)^{\beta}-\left(\frac{t}{\eta}\right)^{\beta}}\left(-\frac{1}{e^{\log_2(N)\left(\frac{t}{\eta}\right)^{\beta}}}+1\right)}{\eta}\right)}{-\left(-\dfrac{1}{e^{\log_2(N)\left(\frac{t}{\eta}\right)^{\beta}}}+1\right)^2+1}
$$

$$(5.65)$$

$$
\lambda_{\text{MCN}} = \frac{-\left(\dfrac{e^{(-N)\left(\frac{t}{\eta}\right)^{\beta}-\left(\frac{t}{\eta}\right)^{\beta}}\left(N\beta\left(\frac{t}{\eta}\right)^{\beta-1}+\beta\left(\frac{t}{\eta}\right)^{\beta-1}\right)}{2\eta} + \dfrac{e^{(-2)N\left(\frac{t}{\eta}\right)^{\beta}+3\left(\frac{t}{\eta}\right)^{\beta}}\left(N\beta\left(\frac{t}{\eta}\right)^{\beta-1}-\frac{3\beta\left(\frac{t}{\eta}\right)^{\beta-1}}{2}\right)}{\eta} - \dfrac{2N\beta\left(\frac{t}{\eta}\right)^{\beta-1}}{\eta e^{N\left(\frac{t}{\eta}\right)^{\beta}}}\right)}{\dfrac{e^{(-N)\left(\frac{t}{\eta}\right)^{\beta}}+e^{-\left(\frac{t}{\eta}\right)^{\beta}}\left(e^{(-2)\left(\frac{t}{\eta}\right)^{\beta}}\left(1-\left(1-e^{(-(N-3))\left(\frac{t}{\eta}\right)^{\beta}}\right)^2\right)\right)+\left(1-e^{-\left(\frac{t}{\eta}\right)^{\beta}}\right)e^{(-N)\left(\frac{t}{\eta}\right)^{\beta}}}{2}}
$$

$$
= \frac{-\dfrac{e^{(-N)\left(\frac{t}{\eta}\right)^{\beta}-\left(\frac{t}{\eta}\right)^{\beta}}\left(N\beta\left(\frac{t}{\eta}\right)^{\beta-1}+\beta\left(\frac{t}{\eta}\right)^{\beta-1}\right)}{2\eta} + \dfrac{e^{(-2)N\left(\frac{t}{\eta}\right)^{\beta}+3\left(\frac{t}{\eta}\right)^{\beta}}\left((-N)\beta\left(\frac{t}{\eta}\right)^{\beta-1}+\frac{3\beta\left(\frac{t}{\eta}\right)^{\beta-1}}{2}\right)}{\eta} + \dfrac{2N\beta\left(\frac{t}{\eta}\right)^{\beta-1}}{\eta e^{N\left(\frac{t}{\eta}\right)^{\beta}}}}{-\dfrac{e^{(-N)\left(\frac{t}{\eta}\right)^{\beta}-\left(\frac{t}{\eta}\right)^{\beta}}}{2} + \dfrac{2}{e^{N\left(\frac{t}{\eta}\right)^{\beta}}} - \dfrac{e^{(-2)N\left(\frac{t}{\eta}\right)^{\beta}+3\left(\frac{t}{\eta}\right)^{\beta}}}{2}}
$$

$$(5.66)$$

$$
\lambda_{\text{network1888}} = \frac{e^{10\left(\frac{t}{\eta}\right)^{\beta}}\left(\dfrac{10\beta\left(\frac{t}{\eta}\right)^{\beta-1}\left(-\left(-\frac{1}{e^{\left(\log_2(N)-1\right)\left(\frac{t}{\eta}\right)^{\beta}}}+1\right)^8+1\right)}{\eta e^{10\left(\frac{t}{\eta}\right)^{\beta}}} + \dfrac{8\beta\left(\log_2(N)-1\right)\left(\frac{t}{\eta}\right)^{\beta-1}e^{\left(-\log_2(N)\right)\left(\frac{t}{\eta}\right)^{\beta}-9\left(\frac{t}{\eta}\right)^{\beta}}\left(-\frac{1}{e^{\left(\log_2(N)-1\right)\left(\frac{t}{\eta}\right)^{\beta}}}+1\right)^7}{\eta}\right)}{-\left(-\dfrac{1}{e^{\left(\log_2(N)-1\right)\left(\frac{t}{\eta}\right)^{\beta}}}+1\right)^8+1}
$$

$$(5.67)$$

5.4.2 Numerical Results

It is necessary to determine the value of the variable parameters in the equations to obtain numerical results from the equations obtained in the Sect. 5.4.1. The variable parameters include the following: Weibull distribution metrics includes: t is lifetime parameter (or operating time), η is the scale parameter (characteristic life), and β is the shape parameter, also known as the Weibull slope. In addition, the size of network (N) is essential.

The shape parameter (slope parameter) β is frequently chosen over a wide range of $\beta < 1$, $\beta = 1$, and $\beta > 1$ [25, 26]. For the value in range of $\beta < 1$, we found that the failure rate (hazard function) is decreased in over time. Significantly, the infant mortality, or untimely failure of defective items, and decreasing failure rates in overtime as defective cases are eliminated from the population.

Moreover, for the value of $\beta = 1$, we found that the failure rate is determined in over time. As a result, random external events may cause mortality or failure. Finally, with the value in range of $\beta > 1$, we found that the failure rate will be increased in time. This happens if there is an "aging" process, or some parts exist that are more likely to fail as time goes on.

Typically, in the field of interconnection networks, increasing in operating time (t) causes the growing of the failure probability of network equipment (as links and switching elements). So, as mentioned, the range of $\beta > 1$ will be a qualified choice for utilizing in interconnection networks. However, according to the graphs in Fig. 5.2, $\beta = 2$ is an acceptable value for this case.

For identifying the correct value of the characteristic life or scale parameter η, it should be considered in Weibull probability density function (Weibull pdf). Commonly, the Weibull pdf provides the relative frequency of failure times in a time function. The scale parameter (η) is 63.2% data. For instance, with 20 for scaling, 63.2% the equipment will be failed in the first 20 h next of the threshold time. Figure 5.3 represents the impact of the scale parameter (η) on the Weibull pdf (with $\beta = 1$ and $\beta = 2.5$). As shown in Fig. 5.3, in early times, with $\beta = 2.5$ and $\eta = 50$, the relative frequency of failure times is upper than $\eta = 100$. Really, when η is increased, the height of pdf curve will be decreased. However, by increasing the η, the relative frequency of failure times has covered more times.

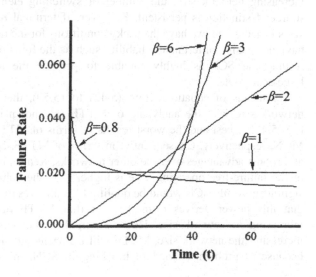

Fig. 5.2 Altered values of β and Weibull failure rate with them

Fig. 5.3 Weibull *pdf* with different values of η

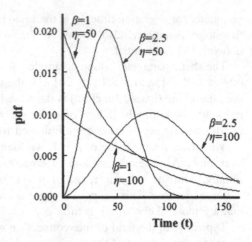

In the field of interconnection networks, it is usually assumed that the relative frequency of failure times will be more distributed at the high operating times. Because of this point and Fig. 5.3, we define an operating time range (*t*) from 1000 to 8000 h and $\eta = 6000$ which is a reasonable estimate for this study. In addition, the results for different network sizes of 8–1024 will be analyzed.

In Tables 5.2, 5.3, and 5.4, the results of analyzing of terminal reliability as a function of time for various network sizes (8, 128, and 1024) are represented briefly.

As revealed, SCN has a significant feature in terminal reliability relative to other networks, e.g. MCN, SEN, SEN+, replicated MIN, Benes network, network 1248, and network 1888. As an important point, SCN can preserve its benefit in larger sizes of network. However, other networks suffer from the problem of decreased terminal reliability in large network sizes. In the SCN, with the condition of increasing network size, the number of switching elements among each pair of source–destination is persistent. However, if terminal reliability in large network sizes is decreased, we have the tricky conditions for the rest of the networks. So, we have decreasing of terminal reliability such as the function for network size. As the results, the SCN is highly capable to provide the terminal reliability yet in large-scale states.

By means of equations from (5.42) to (5.50), there will be the function of network size with the analyzing of MTTF (as shown in Fig. 5.4). According to Fig. 5.4, the best and the worst results in terms of MTTF are owned by SCN and MCN, respectively. Consequently, in terms of MTTF, SCN has relatively definite and robust advantages over the other networks. Really, in SCN, the expected value of the failure-free operating time is higher than the value of other networks. As a significant result, SCN would be qualified for all sizes of network. It can be claimed that this power derives from the fact that MTTF of SCN is not affected by increasing of network size. However, in the other cases of the networks, with increasing the network size, MTTF will be decreased significantly. Consequently, because of problem solving of blocking in SCN, the final results prove that in

Table 5.2 Terminal reliability as a function of the network size 8

Time (h)	SCN	SEN	SEN+	Benes network	Replicated MIN	MCN	Network 1248	Network 1888
1000	0.972604	0.920044	0.943197	0.943124	0.966387	0.863624	0.891952	0.757465
2000	0.894839	0.716531	0.768944	0.766055	0.822935	0.520346	0.60695	0.329192
3000	0.778801	0.472367	0.512629	0.49793	0.561985	0.198584	0.280497	0.082038
4000	0.64118	0.263597	0.268543	0.241762	0.293475	0.046425	0.078586	0.011574
5000	0.499352	0.124514	0.108849	0.084478	0.116611	0.006707	0.012799	0.000867
6000	0.367879	0.049787	0.034153	0.021101	0.116611	0.000608	0.00123	0.000031
7000	0.256376	0.016851	0.008357	0.003806	0.008568	0.000035	0.000072	0.000001
8000	0.169013	0.004828	0.001609	0.000503	0.001628	0.000001	0.000003	3.935912E−9

Table 5.3 Terminal reliability as a function of the network size 128

Time (h)	SCN	SEN	SEN+	Benes network	Replicated MIN	MCN	1248	1888
1000	0.972604	0.823292	0.923665	0.942872	0.942234	0.042796	0.891832	0.757465
2000	0.894839	0.459426	0.611152	0.75151	0.633349	0.000001	0.59358	0.328159
3000	0.778801	0.173774	0.240473	0.392761	0.247153	2.0486E−14	0.21782	0.071194
4000	0.64118	0.044551	0.055146	0.078667	0.055858	1.3353E−25	0.031489	0.005143
5000	0.499 352	0.007 742	0.007672	0.004832	0.007702	1.8676E−39	0.001822	0.000113
6000	0.367879	0.000912	0.00067	0.000111	0.000671	2.0990E−56	0.000049	0.000001
7000	0.256376	0.000073	0.000037	0.000001	0.000037	1.8909E−76	0.000001	2.7841E−9
8000	0.169013	0.000004	0.000001	6.527E−9	0.000001	1.3662E−99	4.34463E−9	3.5464E−12

Table 5.4 Terminal reliability as a function of the network size 1024

Time	SCN	SEN	SEN+	Benes network	Replicated MIN	MCN	1248	1888
1000	0.972604	0.757465	0.899675	0.942872	0.915393	6.7120E−13	0.891548	0.757461
2000	0.894839	0.329193	0.480782	0.751174	0.492178	2.1346E−50	0.562829	0.320801
3000	0.778801	0.082085	0.121118	0.370015	0.122608	4.0398E−112	0.138879	0.048411
4000	0.64118	0.011744	0.014922	0.039696	0.014971	1.5131E−198	0.009635	0.001614
5000	0.499352	0.000964	0.000962	0.000597	0.000962	1.1056E−309	0.000238	0.000015
6000	0.367879	0.000045	0.000033	0.000002	0.000033	1.5637E−445	0.000002	4.4803E−8
7000	0.256376	0.000001	0.000001	2.271E−9	0.000001	4.2699E−606	1.0869E−8	4.6961E−11
8000	0.169013	1.901E−8	6.429E−9	1.217E−12	6.429247E−9	2.2525E−791	2.0934E−11	1.7123E−14

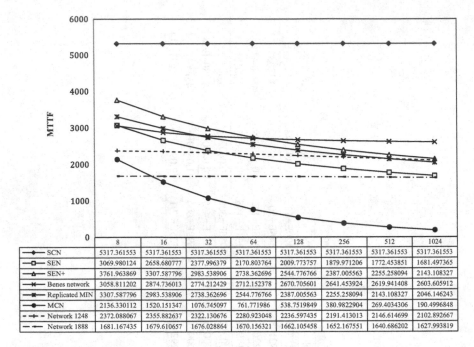

	8	16	32	64	128	256	512	1024
◆ SCN	5317.361553	5317.361553	5317.361553	5317.361553	5317.361553	5317.361553	5317.361553	5317.361553
□ SEN+	3069.980124	2658.680777	2377.996379	2170.803764	2009.773757	1879.971206	1772.453851	1681.497365
△ SEN+	3761.963869	3307.587796	2983.538906	2738.362696	2544.776766	2387.005563	2255.258094	2143.108327
✳ Benes network	3058.811202	2874.736013	2774.212429	2712.152378	2670.705601	2641.453924	2619.941408	2603.605912
■ Replicated MIN	3307.587796	2983.538906	2738.362696	2544.776766	2387.005563	2255.258094	2143.108327	2046.146243
● MCN	2136.330112	1520.151347	1076.745097	761.771986	538.7519849	380.9822904	269.4034306	190.4996848
- + - Network 1248	2372.088067	2355.882637	2322.130676	2280.923048	2236.597435	2191.413013	2146.614699	2102.892667
- ▬ - Network 1888	1681.167435	1679.610657	1676.028864	1670.156321	1662.105458	1652.167551	1640.686202	1627.993819

Fig. 5.4 MTTF as a function of network size

multiprocessor systems, SCN has more qualified options for utilizing comparing to the other networks.

However, according to Eqs. (5.52) through (5.59), the effects of analyzing of the cost-effectiveness (CE) are shown in Fig. 5.5. As it is revealed, the SCN network performance is almost appropriate concerning effectiveness cost, mostly in small-scale network. Furthermore, as it is shown in Fig. 5.5, MCN in terms of effectiveness cost provides low-value results. Really, the efficiency of MCN in respect of cost-effectiveness comparing to SCN is very unexpected, meanwhile SCN and MCN networks would be able to have a non-blocking condition. Also, SCN has the best achievement about network sizes 8 and 16 relation to other networks. Besides, it has the best achievement about network size 32 relation to Benes network, replicated MIN, MCN, network 1248, and network 1888. Additionally, SCN has an important benefit effectively over all network sizes comparing to MCN, network 1248, and network 1888 and regarding to cost-effectiveness. Figure 5.5 is shown results for other states, and SCN has near competition with all the networks in respect of cost-effectiveness. According to aforementioned, we can find that SCN network encounters with important metrics such as scalability, non-blocking condition, reliability, MTTF, and efficient routing mechanism, but the excessive cost cannot be enforced to these vital benefits.

Generally, in each unit time, the failure rate of system can be defined as the failures number for a specific period of time. Thus, at any time the value of this metric is low, the network would be operated, while the number of failures in a

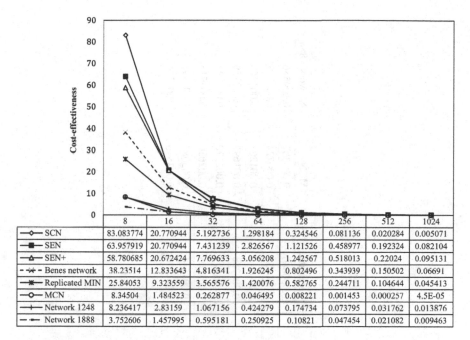

	8	16	32	64	128	256	512	1024
—◇— SCN	83.083774	20.770944	5.192736	1.298184	0.324546	0.081136	0.020284	0.005071
—■— SEN	63.957919	20.770944	7.431239	2.826567	1.121526	0.458977	0.192324	0.082104
—△— SEN+	58.780685	20.672424	7.769633	3.056208	1.242567	0.518013	0.22024	0.095131
- -✳- - Benes network	38.23514	12.833643	4.816341	1.926245	0.802496	0.343939	0.150502	0.06691
—✱— Replicated MIN	25.84053	9.323559	3.565576	1.420076	0.582765	0.244711	0.104644	0.045413
—○— MCN	8.34504	1.484523	0.262877	0.046495	0.008221	0.001453	0.000257	4.5E-05
—+— Network 1248	8.236417	2.83159	1.067156	0.424279	0.174734	0.073795	0.031762	0.013876
— ⋅— Network 1888	3.752606	1.457995	0.595181	0.250925	0.10821	0.047454	0.021082	0.009463

Fig. 5.5 Effectiveness cost such as a function of network size

specific period of time will be decreased. Also, in Eqs. from (5.61) to (5.67), the system failure rate inspection that assumed such as a function of time for different network sizes 8, 128, and 1024 would be entered into Tables 5.5, 5.6, and 5.7, correspondingly. In the tables from 5.5 to 5.7, the consequences of the analyzing of system failure rate are certain for different time durations (from 1000 to 8000 h). According to the formerly findings, SCN regards to other networks has the lowest failure rate in sizes of 8, 128, and 1024. Naturally, the failure rate goes up steadily as the time increases. Nevertheless, failure rate of SCN is totally less between the other networks at all times. Furthermore, MCN network has the lowest performance with respect to failure rate. Actually, these achievements reveal that MCN as SCN can ignore the blocking issue, but when MCN is concerned to failure rate as the important performance metrics, it is not appropriate. Moreover, combining MCN's high complexity in network and high cost of hardware is the main reason of its weak failure rate

In summary, analysis of important performance metrics approves that SCN achieves significant benefits over the blocking SEN, reorganizable Benes network, non-blocking MCN, fault-tolerant SEN+, replicated MIN, and multilayer MIN 1888. SCN eliminates the challenging issue of blocking. Also, it yields a very good performance in terms of terminal reliability, mean time to failure, and system failure rate.

Seeking some notions are necessary in future studies: SCN can be considered with some cutting-edge design concepts to achieve topologies with

Table 5.5 Failure rate of system as a function of time in network size 8

Time (h)	SCN	SEN	SEN+	Benes network	Replicated MIN	MCN	Network 1888
1000	0.000056	0.000167	0.000122	0.000123	0.000080	0.000307	0.000556
2000	0.0001111	0.000333	0.000296	0.000306	0.000258	0.000724	0.001111
3000	0.0001667	0.0005	0.000522	0.000567	0.000512	0.001208	0.001669
4000	0.0002222	0.000667	0.000774	0.000884	0.000788	0.001697	0.002259
5000	0.0002778	0.000833	0.001032	0.001220	0.001056	0.002170	0.002943
6000	0.0003333	0.001	0.001285	0.001552	0.001308	0.002630	0.003713
7000	0.0003889	0.001167	0.001529	0.001871	0.001546	0.003084	0.004495
8000	0.0004444	0.001333	0.001765	0.002175	0.001774	0.003536	0.005246

Table 5.6 Failure rate of system as a function of time in network size 128

Time (h)	SCN	SEN	SEN+	Benes network	Replicated MIN	MCN	Network 1888
1000	0.000056	0.000389	0.000199	0.000125	0.000172	0.007021	0.000556
2000	0.0001111	0.000778	0.000659	0.000366	0.000657	0.014190	0.001129
3000	0.0001667	0.001167	0.001208	0.001037	0.001222	0.021293	0.002018
4000	0.0002222	0.001156	0.001730	0.002217	0.001742	0.028402	0.003244
5000	0.0002778	0.001944	0.002209	0.003317	0.002215	0.035516	0.004355
6000	0.0003333	0.002333	0.002664	0.004217	0.002666	0.042633	0.004745
7000	0.0003889	0.002722	0.003111	0.005645	0.003111	0.049751	0.00622
8000	0.0004444	0.003111	0.003555	0.004701	0.003555	0.056870	0.007111

Table 5.7 Failure rate of system as a function of time for network size 1024

Time (h)	SCN	SEN	SEN+	Benes network	Replicated MIN	MCN	Network 1888
1000	0.000056	0.000556	0.000292	0.000125	0.000272	0.056871	0.000556
2000	0.0001111	0.001111	0.000997	0.000369	0.001003	0.113746	0.001233
3000	0.0001667	0.001667	0.001750	0.001245	0.001762	0.170626	0.00265
4000	0.0002222	0.002222	0.002426	0.003303	0.002431	0.227513	0.004096
5000	0.0002778	0.002778	0.003053	0.004966	0.003054	0.284393	0.005186
6000	0.0003333	0.003333	0.003666	0.006854	0.003667	0.341273	0.006332
7000	0.0003889	0.003889	0.004278	0.008291	0.004278	0.398153	0.007389
8000	0.0004444	0.004444	0.004889	0.006888	0.004889	0.455033	0.008444

higher-performance. For example, multilayer and replicated networks can be regarded with SCN topology. Furthermore, some progressive methods for example adding up to groups of switches, connect the groups of switches to each other, and designing asymmetric networks (networks with different numbers of source and destination nodes) can be inferred in SCN.

References

1. Bansal PK, Joshi RC, Singh Kuldip (1994) On a fault-tolerant multistage interconnection network. Comput Electr Eng 20(4):335–345
2. Birolini A (2014) Reliability engineering: theory and practice. Springer, Berlin Heidelberg
3. Bistouni F, Jahanshahi M (2015) Pars network: a multistage interconnection network with fault-tolerance capability. J Parallel Distrib Comput 75:168–183
4. Cuda D, Giaccone P, Montalto M (2012) Design and control of next generation distribution frames. Comput Netw 56(13):3110–3122
5. Gunawan I (2008) Redundant paths and reliability bounds in gamma networks. Appl Math Model 32(4):588–594
6. Tutsch D, Hommel G (2008) MLMIN: a multicore processor and parallel computer network topology for multicast. Comput Oper Res 35(12):3807–3821
7. Das N, Mukhopadhyaya K, Dattagupta J (2000) O (n) routing in rearrangeable networks. J Syst Architect 46(6):529–542
8. Newman P (1989) Fast packet switching for integrated services. University of Cambridge, Computer Laboratory
9. Bistouni F, Jahanshahi M (2014) Improved extra group network: a new fault-tolerant multistage interconnection network. J Supercomput 69(1):161–199
10. Kolias C, Tomkos I (2005) Switch fabrics. IEEE Circuits and Devices Mag 21(5):12–17
11. Bauer E (2010) Design for reliability: information and computer-based systems. Wiley, Hoboken, New Jersey
12. Bistouni F, Jahanshahi M (2014) Analyzing the reliability of shuffle-exchange networks using reliability block diagrams. Reliab Eng Syst Saf 132:97–106
13. Blake JT, Trivedi KS (1989) Reliability analysis of interconnection networks using hierarchical composition. IEEE Trans Reliab 38(1):111–120
14. Blake JT, Trivedi KS (1989) Multistage interconnection network reliability. IEEE Trans Comput 38(11):1600–1604
15. Dash RK et al (2012) Network reliability optimization problem of interconnection network under node-edge failure model. Appl Soft Comput 12(8):2322–2328
16. Fard NS, Gunawan I (2002) Reliability bounds for large multistage interconnection networks. Appl Parallel Comput. Springer, Berlin Heidelberg
17. Gunawan I (2008) Reliability analysis of shuffle-exchange network systems. Reliab Eng Syst Saf 93(2):271–276
18. Kumar VP, Reddy SM (1987) Augmented shuffle-exchange multistage interconnection networks. Computer 20(6):30–40
19. Lee SE (2013) Adaptive error correction in orthogonal latin square codes for low-power, resilient on-chip interconnection network. Microelectron Reliab 53(3):509–511
20. Mettas A, Savva M (2001) System reliability analysis: the advantages of using analytical methods to analyze non-repairable systems. In: Reliability and maintainability symposium, 2001. Proceedings. Annual. IEEE
21. Veglis A, Pomportsis A (2001) Dependability evaluation of interconnection networks. Comput Electr Eng 27(3):239–263

22. Zhu Q, Wang X-K, Cheng G (2013) Reliability evaluation of BC networks. IEEE Trans Comput 62(11):2337–2340
23. Wei S, Lee G (1988) Extra group network: a cost-effective fault-tolerant multistage interconnection network. In: ACM SIGARCH computer architecture news, vol 16, no 2. IEEE Computer Society Press
24. Koren I, Mani Krishna C (2007) Fault-tolerant systems. Morgan Kaufmann, USA
25. McCool JI (2012) Using the Weibull distribution: reliability, modeling and inference. Wiley, Hoboken, New Jersey
26. Stapelberg RF (2009) Handbook of reliability, availability, maintainability and safety in engineering design. Springer, London

Index

Printed in the United States
By Bookmasters